Backyard
BEES

Backyard BEES

A guide for the beginner beekeeper

DOUG PURDIE

CONTENTS

INTRODUCTION

I never set out to become a beekeeper; the bees found me and my life has been different ever since. I started my working life cooking chickens at a supermarket in Fremantle, Western Australia. My family, being part Italian, has always had a strong connection with food and where it comes from, so it was a strange place to begin. My grandmother and aunts used to make all their own cheese, wine, vinegar (sometimes it was hard to tell the wine from the vinegar) and smallgoods. They grew pretty much all their own greens, had about 30 chickens, kept pigeons and were largely self-sustained. I grew up eating amazing food.

I moved on from chickens and studied cinematography, then worked for a large film lab. Later I worked in news and current affairs at a TV station, and finally got into IT and marketing. Along the way I discovered that I could build and fix pretty much anything—and that I loved food.

In 2009 or thereabouts I was reading about food trends (I had just started reviving the cheese-making of my family's past) when I read about clandestine rooftop beekeeping in New York and the massive problems bees face in most of the world due to widespread diseases. Wow, I thought, why don't I know about this?

I realised that here in Australia nobody seemed to know that bees were really in trouble. I decided to do something about it and started reading book after book on beekeeping.

I had been selling honey on behalf of my local community association at a farmers' market for some time with no interest in actually having bees of my own, but as I read more and more I became hooked, and was eventually given my first hive by my then father-in-law. It was a baptism of fire as it was a most badly behaved hive with very grumpy bees—even the fearless Ian from the local bee club wouldn't handle the bees unless he was fully kitted up.

I also joined the Amateur Beekeepers' Association of New South Wales Sutherland branch, which at the time was the closest to my home, some 45 minutes drive away. My first visit to the club revealed a room full of older, mainly male members, presided over by a man with a gavel who was hard of hearing. What was immediately apparent was the huge amount of knowledge these people had, and as they were not getting any younger I decided to get involved and save the knowledge before it disappeared.

From there I became the president of the NSW association, a position that had been empty for some years, and started banging on about bees to anybody who would listen.

I met my business partner Vicky via the association and we formed The Urban Beehive, our aim being to place beehives in community gardens, backyards and cafés. Our first commercial beehive installation was at the Swissotel in Sydney's CBD; we nervously wheeled four beehives full of stinging insects through the hotel and installed them near the pool on the roof. They were a success and we now have a growing number of sites and hives, and beekeeping is a large part of my day-to-day life.

We started the Sydney City Branch of the Amateur Beekeepers' Association in 2012 and it's growing rapidly, with a lot of much younger members than the other branches. Many of our members have backyards with chickens and vegetable gardens and see bees as just another produce.

In my beevangelist work (as I like to call it) I find myself in fairly constant demand talking to a growing sphere of people via all sorts of media, from my blog to radio, newspaper and TV—even taking part in a TV cooking segment.

One of the unbelievable things that I have discovered talking to people about bees is that we are so disconnected from our food. Many people don't even realise that bees pollinate our food and that without them we have a major problem ... and a very boring diet of grain.

Bees have a bad rap and I would like to try to educate people about how easy bees are to keep; they don't take a lot of time and by having a hive you are helping the planet, and your neighbour's vegetable patch.

Beekeeping is now the new black; lots of people are giving it a try or have been given a beehive as a gift—I receive a number of calls at Christmas from people wanting to buy beehives—and need guidance. I have read just about every bee book there is and they are often so dry and preachy that no bearded hipster is going to pick them up.

Hopefully the bee stories in this book will inspire some enthusiastic new beekeepers, who in turn will pass on their skills, continuing the time-honoured beekeeping tradition.

Opposite: The varied colours of honey are indicative of bees foraging from different nectar sources.

1

WHY KEEP BEES?

Bees are incredible insects. They existed before the dinosaurs and evolved at about the same time as flowers in order to pollinate them—thousands of years ago. Only now have we worked out how to wipe them out. Bees are under threat all around the world, with massive winter bee die-offs reported from almost every place and insecticide bans enforced in many countries to try to stem the losses.

Australians are very lucky. At the time of writing Australia is the only country in the world without varroa mite and colony collapse disorder—very lucky indeed. But because of this good fortune Australians are complacent and don't care for their bees like they should. During spring when the bees are swarming the most common reaction is to kill them, even from public officials.

Many people don't realise the vital role bees and other insects play in pollinating our food. When bees forage for nectar and pollen they pollinate each flower they visit, increasing the crop yield by as much as 60 per cent. Without bees, many food crops that need pollination by insects couldn't be grown on a scale large enough to feed us. In some parts of the world, plants have to be hand-pollinated, a very labour-intensive and expensive process where a human touches each flower with a feather.

Previous page: Fully capped frame of honey ready for harvesting. Opposite: Brood comb, showing the natural comb shape bees make in nature.

In the apple and pear orchards of China's south-west, bees have been eradicated by pesticide use and habitat loss, and hand pollination is the only option.

This is the reason I started beekeeping: reading about bee decline and its impact, and realising that in Australia most people had not heard of the problems the rest of the world is facing. I wanted to get involved and tell people that bees need saving.

A few years on the message is getting out and people are starting to listen. You too can help by telling everybody you know about bees and how we need them to pollinate our plants for food and other crops.

As well as the benefit to bee numbers, beekeeping is an amazing and rewarding hobby whether you eat lots of honey or not. Even after years of keeping bees I find them fascinating. If you get your own hive you will spend a lot of time just watching them come and go on their mission to pollinate and collect pollen and nectar. Sitting beside your hive with a cup of tea watching the girls (as I fondly refer to them) come and go is very gratifying; imagining where they are going and trying to identify the trees they have visited from the colour of the pollen sacks on their hind legs.

The first time you open a hive it's overwhelming. There are lots of bees buzzing around all armed with the much-feared stinger. But you soon realise that they are ignoring you and simply going about their business, seemingly unaware of the intruder in their midst.

Then there is that special moment of seeing a bee being born for the first time, as she crawls out of her cell all fluffy and new and ready to start work. Or seeing a queen being born—that is something special indeed.

So get a good grip on your fear of bees, if you have one, and get involved. You will be rewarded with a lovely gift: the gift of interacting with a super organism—a beehive. Plus, you will be giving them a helping hand in surviving the insect apocalypse that is upon us. At the same time, tell people about what you are doing and why they should re-think that green lawn and plant flowers instead. And put down that insecticide can.

15

Opposite: Two Langstroth hives placed side by side. The different hive colours help the bees locate their hive.

Which bees are for you?

It's a question I get asked a lot. Shouldn't you just keep native bees? Aren't the other bees introduced and bad for the environment? This is a complex question and many native bee species are under threat from over-development and loss of habitat, along with pesticide use. No matter where you live in the world there are many native bee species that need our care. We also need nature's super pollinator, the European honey bee, as there really is no better pollinator for the introduced plant species that are our food supply. It is my belief that all bee species need our support. There is little evidence to suggest that different bee species do not tolerate each other and, with the exception of a few predatory species, most bees happily coexist. In the urban environment there is so much forage available that having European honey bees is unlikely to be taking food from native bees.

You can help native bees by providing habitat in the form of bee houses or, in the case of social bees, perhaps purchasing and caring for a hive of them. The social native bee species in Australia is *Tetragonula*, which typically builds nests in old trees and has a limited foraging range of 500–800 metres (545–875 yards). These 'stingless' bees are very susceptible to fluctuating temperatures and need to be located with care to make sure they have enough food and are not chilled in winter or overheated in summer. They are fascinating to watch and make a great addition to your backyard. They are not for honey production though and I am of the view that you should let them be and not try to take their honey unless you live in a very warm, productive location where they will produce an excess.

Harvesting of wild honey has been practised for many thousands of years and is still practised today by Aboriginal societies in Australia, Africa, Asia and South America. It's fraught with danger, involving climbing to great heights with no protective equipment and a small fire to produce smoke. Experienced harvesters have often been blinded by the stings; not my idea of honey harvesting, I can tell you.

European honey bees also have stings but don't let that deter you. These bees have been kept since Egyptian times, as ancient inscriptions show. There was even a medieval practice of migratory beekeeping: floating hives down the Nile. The majority of medieval beekeeping (by which I mean putting them in a hive of some sort and harvesting the honey) was focused around abbeys and monasteries, mainly for producing beeswax for candles. The honey was often used to make mead. The downside was that they were using skep hives, the hives made of wicker or straw and mud that most people today associate with honey. The bees attached their honeycomb to the interior of the hive, making it difficult to harvest without destroying the whole hive.

So while I like native bees I only manage European honey bee hives and this book covers keeping *Apis Mellifera* primarily in a Langstroth hive. Reverend Lorenzo Langstroth, a nineteenth-century apiarist, teacher and clergyman who is considered to be the father of American beekeeping, first described this type of hive in 1853. Langstroth was himself referencing work by another pioneering apiarist, the blind Swiss naturalist Francis Huber, who in 1789 invented the first hive with moveable frames.

Langstroth is famous for first describing 'bee space', which is based on the notion that if bees are given less than 1 centimetre (½ inch) of space, they will not fill the hive with wax or propolis, the resin bees collect from trees and use as a sealant. Langstroth designed a hive that gave 6–8 millimetres (¼–⅜ inch) of space between all the components of the hive, meaning the bees would not seal it shut with propolis, allowing an apiarist to remove honey frames without destroying the hive and killing the colony.

I use the Langstroth hive mostly because it's easily available and bees are quite happy in it. In fact bees really don't care what sort of hive you use; they are equally happy in a barbecue lid or a compost bin. It's the method and attitude of the beekeeper that matters to bees, not the box the bees call home. Beekeeping needs to be easy for the beekeeper and the bees as bees need looking after and will not prosper if neglected. I don't believe in tying yourself in knots about the type of hive you use; instead, think about not using chemicals and caring for your girls in an organic manner.

Rather than one hive I suggest you start with two hives. This enables you to compare the hives if something seems not quite right, and should a queen falter you can help the ailing hive by moving a frame with eggs across so they can make a new queen (see page 134; Queen problems and re-queening).

17

Bee stings and allergies

I must have heard 'but I'm allergic to bees' thousands of times.
In fact, almost everybody swells up when they are stung; it's the
normal reaction to a bee sting. Most beekeepers are stung many
times and their body gets used to the sting so they no longer
swell up, though they still hurt. Some stings hurt much more than
others—I like to say that some bees mean it more. Sometimes I
swell up a little and sometimes a lot.

When I was a new beekeeper my first hive was so angry I
could hardly see out of my veil for the number of grumpy bees
buzzing about my head. Unfortunately, my girlfriend at the time,
who was minding her own business some 150 metres (490 feet)
from the hive, was singled out and stung on the forehead. It was
not a particularly painful sting but at about 2 am she woke me
to say her face felt funny. I turned on the light to be greeted by
a horribly swollen closed eye; it looked like she'd been in a fight.
Happily, the swelling soon went down.

Stings are all part of the game—if you keep bees it's going to
happen. Having said that, some beekeepers can become allergic
to the stings over time so it's best not to be too brave and avoid
getting stung where you can by using protective gear. If you
do get stung through your clothing, give the area a little puff of
smoke to mask the smell of the sting and prevent further stings
to the same area.

According to an Australian coroner's report on beekeeping
in urban areas, conducted as a result of the death of a woman in
2000 after her mother's bees attacked, 'The risk of a fatal allergic
response to a bee sting is very, very low. Allergies to penicillin
cause more deaths than bee venom and individuals in Australia
have a greater likelihood of being killed by lightning than by
bee venom.'

Unfortunately, this information is probably not going to calm
the average neighbour who is sure your bees are going to attack
them at the first opportunity. Once the bees start producing,
dropping by with a jar of honey often helps quell their fears.

19

Opposite: A beekeeper
holding a brood
frame and wearing full
protective gear: boots,
suit and gloves.

Myths and facts

Bee stings are good for arthritis.
I know plenty of beekeepers with arthritis so I'm not sure about this one.

Honey is antibiotic.
This is fact. Honey—and I'm not just talking about so-called active honey like the famed Manuka—can be used to treat all sorts of infection. Pretty much all honey has some antibiotic activity, with tea tree (*Leptospermum*) honey being the best. It tastes pretty average though.

Bees get to know their keeper.
This is not true. Bees have very short lives of about 40 days and will have forgotten who you are even between hive inspections.

Honey can be diluted with water or sugar before it's sold.
It's true that you can dilute honey with water, but it will ferment and smell disgusting, and it can no longer be sold as honey. Candied honey—honey that has gone hard and has a white-ish tinge—has not had sugar added; it's just raw honey that's undergone a normal reaction.

Bees are naturally aggressive.
Not true. Bees will not seek you out unless they believe you are threatening their hive, and even then the bees we keep these days are bred to be passive. The cartoon vision of bees rushing out of the hive to attack is just that, a cartoon—unless you poke the beehive with a stick, and then you deserve it! Just like all wild creatures bees will respond to environmental stresses and their temperament will alter. You also can get genetic variations that produce angrier bees, like Africanised bees—now they are grumpy bees!

Opposite: Lighting the smoker. Cool smoke is used sparingly to help calm the bees.

Ian

A long-standing and devoted member of the Sutherland Bee Club in Sydney, Ian almost never wears protective gear and for some reason bees rarely sting him, despite his long hair and ponytail, something that bees usually don't like.

Ian no longer has hives of his own but that doesn't stop him being involved with bees on a regular basis, conducting tours at the club and performing duties as a bee fixer. He often gets called to help out with sticky bee situations, such as enormous feral beehives in trees, or misbehaving grumpy bees that need re-queening.

2

HOW TO START

Previous page: Electric heated uncapping knife and cappings. Removing the wax caps allows the honey to flow out. Opposite: A hive inspection should be done regularly and only takes about 30 minutes. Light your smoker and inspect each comb, checking the bees' health.

Almost anybody can keep bees. You don't need a huge amount of space—my house in inner-city Sydney is only about 3 metres (10 feet) wide and I've kept bees on the back deck and roof on and off for years. Usually neighbours will be unaware you even have a hive.

Beekeeping is an incredibly rewarding hobby that only takes an average of 40 minutes every two weeks during spring, summer and autumn. Unlike most household animals, the girls feed and water themselves. And then there is the honey. Even if you have a serious honey addiction you will still probably make more than you can eat, leaving excess for your friends and family.

This book is designed to give you all the basic information you need to get going. If you're like me, you will probably want more and more information once you have a taste of beekeeping, and there are plenty of excellent blogs and other online resources that you can read at your leisure.

If the more personal touch is what you're looking for, consider joining your local bee club for access to a wealth of knowledge. You get to ask questions of many beekeepers, all of whom will have a different opinion and probably disagree with each other, but you'll learn a lot in the process! Clubs can also assist with buying bees and gear.

Courses and local clubs

Many clubs also run beekeeping courses. If you are going to do a course—they can be one-day or weekend courses, or longer for professional training—make sure the course includes hands-on experience with bees. Sounds simple but some courses do not include actually handling bees and in my opinion that is not a bee course, that is a bee lecture. So if it's not clear in the sales pitch, ask if you get to handle bees before booking the course.

Once you start talking about bees many people will offer an opinion about a particular hive design. Ignore what anybody tells you about bees preferring this or that, or about this design being cruel and that design being more bee-friendly. Just make up your own mind by doing some research, and recognise that you are the biggest influence on how comfortable the bees are in any type of hive. Any hive design can be used as either a top bar or a foundationless hive (more on that on page 67; Frames and foundation) and really, the bees prefer shelter from the elements first and foremost; everything else is secondary.

Basic beekeeping equipment

Once you've had a go handling bees, know what to expect when you open a hive and can light a smoker, you're ready to get some beekeeping gear. Your basic equipment list will look like this:

1. Woodware—beehive brood box, base, lid and frames
2. Tools—bee smoker, hive tool and brush
3. Gear—personal protective equipment

Woodware

Your basic woodware starter kit consists of: a base; one 'super', the box where bees store honey, to act as a brood box for developing bees; frames to fill the box; and a lid. You will add more supers with frames as the hive grows.

Much of the woodware you purchase will have the option to be bought assembled or unassembled. Unless you are handy with a hammer, I would purchase assembled gear as it's not that much more expensive and assembly must be done with precision. If you decide to buy your hive unassembled, see page 68, Assembling woodware, for basic instructions on how to put your hive together.

The material for your woodware will depend on where you live. In the United States, cypress (particularly in the southern states), cedar and pine are commonly used. In Australia, most of the woodware is made out of radiata pine, which, although a plantation timber and therefore ecologically quite sound, really has no longevity unless painted. I would suggest painting the outside of the boxes with a water-based paint. Any light colour will be fine; avoid dark colours as they can increase the internal temperature by absorbing heat. I often buy mis-tint paint from the hardware store as it's cheaper.

Purists will argue that you shouldn't paint the box. I have tried both—with and without paint—and found unpainted timber will warp in about every two boxes, so for me painting or treating with wax dipping or oil dipping is essential. I also do not use any timber preservative like copper napthenate, though many beekeepers in Australia do, partly because the timber most of our hives are made of is susceptible to fungal rot.

If you do use copper napthenate, you can choose to seal the wood by painting both the inside and outside of the boxes or just the outside. If you paint only the outside the bees will quite quickly coat the inside of the box with a fine layer of wax and propolis, which can contain disease. This is why some people insist you should paint the inside: it seals up cracks that could otherwise harbour American foulbrood spores. Really though, it's up to you.

A note on second-hand woodware: I would not purchase second-hand beekeeping equipment unless you have the budget to have it irradiated (sterilisation performed by specialist companies) before you use it. The risk of disease is just too great so go for new woodwork—the extra outlay is worth it for the peace of mind.

Tools

When it comes to the smoker and hive tool, buy the most expensive smoker you can afford—it's a tool you will use for a long, long time. My first smoker was a cheap one and when I replaced it with a locally manufactured more expensive smoker I was amazed at the difference it made.

The smoker is used to mask the pheromones that bees use to communicate alarm. It also suggests that there is a fire nearby

29

and the bees prepare by filling up on honey in case they need to evacuate the hive, which means they are busy doing other things rather than watching you. You only need a little smoke; there is no need to engulf the hive, just a puff or two is all that's needed.

The hive tool is a matter of personal preference. I use a J-style tool and would not use anything else, while other beekeepers swear that hive tools that look like crowbars are better. The hive tool is a multi-purpose tool used to lever things apart and scrape wax and propolis off surfaces of the hive. In Australia some races of bees produce large amounts of propolis (bee glue made from tree sap) and I find the J-style gives me excellent leverage to get the stuck-together components apart.

When it comes to a bee brush, go for natural rather than artificial bristles. Any style will be fine as long as it is a specific bee brush consisting of a single row of bristles. Don't be tempted to use a dustpan brush or the like—it will annoy the bees, and you do not want annoyed bees.

Gear

Often beekeepers wear less and less protective gear as they get more used to handling their girls, so don't go out and spend $400 on a bee suit. The average backyard beekeeper can get away with those synthetic fabric coveralls you see on crime shows and a simple head-and-shoulders veil. Rubber washing-up gloves will do for your hands, and plastic-sided workman's boots or rubber gardening boots will protect your ankles and feet.

If you spend more than about 30 minutes on hive inspections per fortnight, or if you want to spend a bit more on an upgrade, I would buy beekeeping gloves first, then purchase a full suit. The suit will be cooler than the plastic coveralls and will always be useful to keep wax, honey and propolis off your clothes, saving you ruining another pair of good jeans with bee glue. Whatever you use as protective gear, make sure everything is tucked in and bee-proof—you don't want the little critters crawling up your trouser leg looking for their hive.

Remember, if you do get stung through your clothing, a little puff of smoke to mask the pheromone of the sting will prevent further stings to the same area.

Now that you have a hive, tools and protective gear, let's get some bees.

Opposite: A smoker is used to produce cool smoke to pacify the bees. A good quality smoker will last a long time.

Getting your bees

You have a few options for obtaining bees: collecting a swarm; purchasing a bee package; purchasing a nucleus hive; and buying a second-hand hive.

The first two options can be used for all hive styles as the bees have not established on any frames (and therefore you don't have to accommodate a different style with some sort of conversion board).

The second two options will only work if you are purchasing bees that have been established on the size and style of hive you intend to populate.

Your local bee club can help you find a beekeeping equipment store near you, or they may even operate their own store and often bee club members get a discount.

A bee swarm

Swarms are a great way of obtaining bees: they are free if you can convince a beekeeper to give you a hand or you've watched enough YouTube videos to think you can do it yourself. It's not difficult to catch a swarm and once you have done it once you will wonder what all the fuss is about. If the idea of relying on a video alone is too much, your local club can probably help, and may even run training sessions on swarm catching.

You could also set up your hive as a trap by baiting it with lemongrass oil—with any luck, a swarm will move in during spring.

Remarkably, swarm traps do work. I have caught swarms by using an old 8-frame beehive with no foundation—old equipment works best as it retains the bee smell—and baiting it with a little lemongrass oil. If you are interested, there are a couple of excellent books on this topic, particularly *Swarm Traps and Bait Hives* by McCartney Taylor.

One important point about swarms: you should always provide new, unused wax for the swarm to build on, never drawn comb (drawn comb is wax that bees have already used for storing honey or raising brood). This is because the bees gorge themselves with honey when they decide to swarm and the honey could contain disease. If they have existing comb provided for them they can store the diseased honey in that, when what you really want is for them to use all of that diseased

Opposite (clockwise from top left): Beekeeper George Schwartz in home-made protective gear; a bee swarm; beekeeper Elke Hague; queen mailing cages used to mail queen bees via the post from a breeder.

How to start

honey—by making wax—first. In that way any disease in the honey will not be stored in your hive.

If you do catch a swarm, keep an eye on the queen. She may be old or she may be a virgin; you don't have any way of telling. Because of this some people replace the queens in swarms as a matter of course. I prefer to wait and see what happens. Don't be in too much of a hurry even if you don't see eggs straight away— she might need to mate first. As in all beekeeping, patience is paramount.

A bee package

Beekeeping equipment shops sell package bees in spring—check with your local store if you need to order them. They consist of a queen bee in a cage and 1 or 1.5 kilograms (2 or 3 lb) of bees (8,000 or 12,000 individual bees) in a special box with sugar syrup in a special feeder. You pick them up, take them home and pour them into your hive: it sounds simple and it is. A bee package will ensure you get healthy bees and a young queen, a great way to populate a hive.

The queen cage is placed in the hive between brood frames or under the lid and the bees will release her over a few days while they are preparing the wax comb for her to lay in. Ask the shop for specific instructions about your package.

A nucleus hive

A nucleus hive consists of four or five frames of brood (basically baby bees) and honey, a queen and bees. It is essentially a small beehive that is easily transferred into your own set-up and will grow very quickly into a full-blown beehive. Remember that the frames have to match your hive.

A second-hand hive

Second-hand hives can be purchased from online auction sites or various sellers who advertise online or in magazines. You need to be careful though as second-hand equipment can be diseased or in poor condition, making it virtually worthless. You need an experienced beekeeper to help you assess the hive's condition prior to purchase, and if the seller won't agree then walk away.

I was once called to the home of a retired invalid who had sold his bees at night-time, with the bees sealed in the hive. When the purchaser picked up the hive he discovered the woodware was rotten and bees were escaping, so he just left it in the driveway, never to return. The owner now couldn't get out of his driveway and the bees were very angry and only partially sealed in—a real mess. I had to re-hive the bees in the middle of the driveway and then come back and get them that night—not my idea of fun.

Matt & Vanessa

These two passionate beekeepers left their jobs in IT to start Melbourne City Rooftop Honey, which they began in order to help bring pollinators back to Melbourne. Their aim is to be part of a global effort to help save honey bees from extinction.

Melbourne City Rooftop Honey seeks to address issues of sustainability, as well as to raise awareness of the importance of bees. By placing hives on the roof spaces of cafes, restaurants, hotels and in private gardens, they have reduced the distance from production to plate to mere footsteps.

3

THE LIFE OF BEES

Previous page: Regular brood inspections are part of keeping bees. Here the beekeeper is looking at a brood frame. Opposite (clockwise from top left): Worker bees make up the majority of the population in the hive; drones are easily recognised by their huge eyes; a queen bee, which the beekeeper has marked with a dot of posca or paint marker pen, making her easier to find.

W hen most people think of bees, they think of the queen bee and imagine her running the hive as an all-powerful monarch. In reality, nothing could be further from the truth. The queen is an egg-laying machine that is cared for and protected by the hive as a super organism. But she's not in charge at all.

When you open a beehive you'll find three kinds of bee:

> Worker
> Drone
> Queen

Worker bees

The worker bees are all female and do all the work around the hive, from cleaning and feeding to guard duty and foraging for nectar, pollen and water. They work until they literally wear themselves out and can't fly anymore. Many die outside of the hive, unable to fly home as their wings are so tattered. Their journey from birth to death takes about four to five weeks and during that time they each make roughly a quarter of a teaspoon of honey. Worker bees forage in a 5–8 kilometre (3–5 mile) radius from

Development calendar of bees

Bee Type	Egg	Larva	Cell Capped	Pupa	Average Development Period	Start of Fertility
Queen	3 days	5½ days	7½ days	8 days	16 days	Approx. 23 days
Worker	3 days	6 days	9 days	12 days	21 days (Range: 18–22 days)	N/A
Drone	3 days	6½ days	10 days	14½ days	24 days	Approx. 38 days

the hive, covering a very large area. They are constantly on the lookout for good nectar sources. When they find one they return to the hive and, using a dance, communicate the location of the nectar to the other bees and share a small taste of it so they can find it. You can demonstrate this behaviour by placing some sugar syrup on a saucer in your backyard and watch as, first one bee visits, then two, then three . . . the numbers keep building. Their navigation is so accurate that if you move the saucer a metre to the left or right, the bees will visit exactly where the saucer was, not where you have moved it to.

During a bee swarm while the bees are looking for a new home, the same dance is used to locate and provide a map to the new home. The bees will find that hole in a tree several kilometres away—it's better than any GPS.

Worker bees are very furry when they are first born, the hairs slowly wearing off as they age. Their very first act as a bee is to clean the cell they were born in, and they don't stop working from those first minutes of their lives until the end.

Drones

The drones are the male bees whose only job seemingly is to find a queen and mate with her. They don't do any work around the house—they can't even feed themselves; the worker bees do that. Whenever I explain this to a group of people the men all start smiling and the women nod at this analogy of domestic life.

Except there is a sting in the tail. Come winter there are no queens to mate with and the drones are doing nothing but eating the food. So they get their wings chewed off and are kicked out of the hive. If you observe a hive during late autumn you will see worker bees dragging drones and depositing them unceremoniously on the ground, where they're left crawling with ripped wings, unable to return. Even worse, when they do find a queen and mate with her, their genitals are ripped off and they die. Most men have stopped smiling at this point.

Drones come from unfertilised eggs that the queen deliberately lays for that purpose, and therefore have no fathers, just grandfathers. It might be a bit hard to get your head around but that's how it works.

43

Opposite: The different stages of bee development: eggs are tiny and can only be seen by looking carefully; larvae of different sizes come next; and capped brood is the final stage.

Drones are large, noisy bees with huge eyes and no sting. You can spot them easily in a hive if you don't hear them first. They spend their life—drones have a life span of about 40 days—congregating with other drones, looking out for a queen to mate with. When they see one they all take off trying to catch her.

Queen bees

Queen bees are a marvel. They start life as a regular fertilised egg that would normally turn into a worker bee, except they're fed royal jelly for their whole development. The royal jelly diet triggers an epigenetic response and creates a queen. Queen bees mate at the beginning of their lives and will mate with a number of drones at that time. They then store and use the semen in the reverse order that they received it. When you open a beehive you may notice that some of the worker bees are a little different and that's the different semen causing slight changes in the bees.

The queen bee emits a pheromone that is the natural scent of the hive and tells the bees that this is their home. The guard bees that stand sentry at the entrance to beehives are actually testing for this scent and will eject any bee that does not smell right, unless they are full of nectar in which case they might be admitted. As the queen ages her ability to produce the pheromone reduces, giving the bees the cue to replace her. Queens last as long as their semen stores remain viable, which can be from one to five years depending on how well they mated.

Aged queens are a big problem for bees. The workers need a young fertilised egg or larvae to produce a new queen but as their queen gets older she starts to run out of semen, which causes more unfertilised eggs to be laid. These turn into drones and it's at this point that you'll see your brood frame with drones spread out where you would expect to see the regular brood pattern. When this happens and if the bees haven't noticed and replaced the queen themselves, without intervention the hive is doomed. The beekeeper needs to add either a frame with eggs so that a new queen can be raised or introduce a new queen bee.

Some beekeepers replace their queen bees every year to keep a big, healthy population of bees. Others prefer to let nature take its course. I can see both sides of this argument and only replace failing queens or queens whose progeny have become aggressive.

44

Opposite: Workers surround the queen. She can't feed herself and is reliant on attendants to meet her needs.

Go to page 134; Queen problems and re-queening, for more information about queen issues.

When you order a queen bee she is delivered by mail in a special cage, usually by a very nervous postman.

The hive

The colony is very organised and everybody has tasks to do, the most important job being to maintain the hive temperature at around 35°C (95°F). The worker bees control the temperature in the hive by using their wing muscles to generate heat and cooling through ventilation. By spraying water and evaporating it (a form of air-conditioning), they also control the humidity to within strict levels, necessary to keep the brood and queen healthy. Bee colonies are a regenerating organism that will last indefinitely unless a catastrophe occurs that kills the queen and no new queen is raised, or she dies while mating due to a predator such as a bird, or is otherwise lost.

Inside the hive there are distinct levels that you will come to recognise. Generally, at the top of the hive there will be honey stored in an arch and recognisable by its white cap; the white cap is wax that the bees have used to close up honeycomb cells full of honey. Under the honey there is often an arch of stored pollen mixed with honey, often called bee bread and easily recognised by its multi colours. Under all the food will be the brood, or developing bees. Here you will find various stages of bee development, from eggs to larvae to 'capped brood', the last stage of development and easily recognised by its tan or light brown velvety breathable cap—see the development calendar on page 42.

46

Opposite: A worker bee emerging from her cell. Young bees are incredibly furry and start work immediately.

Vicky

Vicky started beekeeping 15 years ago on Kangaroo Island, a bee sanctuary and home to pure Ligurian bees. Keeping bees in an urban environment has been a long-time dream for Vicky and she now manages apiaries all across Sydney as one half of The Urban Beehive. She enjoys teaching beekeeping courses and giving talks to raise awareness about earth's major pollinators.

Vicky has visited beekeepers in all sorts of locations, locally and around the world, and enjoys being part of a growing international community of passionate apiarists dedicated to saving our pollinators.

4

WHERE TO PUT YOUR HIVES

I hear it a lot: 'I can't have bees, I've nowhere to put them.' I live in a 3-metre (10-foot) wide terrace house and have bees; my neighbour George lives in a similar-sized house and has kept bees there for more than 30 years. Most neighbours would not even be aware that the bees are there unless they see a swarm, or wonder why you've taken up solo fencing as they watch you navigate your rooftop or backyard in your bee suit. Rooftops are often ideal places for bees as they are elevated, which keeps the bees away from people, and once the bees' flight path is elevated they rarely fly back down again.

Bees have long been part of our society. Most of the strains kept today are very passive and can easily be accommodated in urban environments with little or no trouble.

Check with your local council though—many councils have no controls but in some parts of the world, unbelievably, urban beekeeping is banned. If the council says no, ask your bee club about rules as the council representative could be wrong.

Just remember; keep nice bees. Angry bees have no place in an urban setting and just increase the likelihood of a problem. Purchasing a bee package or being able to inspect the bees you are buying will help ensure they are passive bees.

Previous page: A frame of honey. It's not yet ready to harvest as the cells are not fully capped.

Hive location guidelines

There are a few guidelines that are worth considering when choosing your hive site. Keep in mind that hives cannot easily be moved because it upsets the bees' navigation system, explained further on page 128; Moving your hive. So choose the spot for your hive carefully, or enjoy the pain of slowly moving a beehive across your property 1 metre (3 feet) at a time (and you can be guaranteed it will need to traverse the most inconvenient part of your yard).

Here are the key things to consider when choosing your hive location.

1   Hives should be kept away from pedestrian or vehicle traffic and their flight path should not cross footpaths or roads unless they are forced to fly high by screening or hedges. Imagine a motorbike rider trying to remove an errant bee from their helmet, or a driver grappling with a bee that's flown through the car window (numerous insurance advertisements are based on this, I'm sure).

2   Keep the hive out of sight. If the hive is invisible, then nobody will complain. I once had somebody make a complaint about a hive, claiming that the bees swarmed all over him. I asked him exactly what had happened, and when. Trouble was, the hive hadn't yet been installed when he claimed the incident occurred. If you can hide your hive, passers-by remain totally oblivious and complaints are less likely.

3   Always have a water source available for your bees when it's hot. Bees need water and they shouldn't have to rely on Fido's water bowl or your neighbour's swimming pool for hydration. A water source can consist of a simple bird bath with a few floating corks for the bees to land on, or a more sophisticated arrangement with a sloping gravel bed and float valve to keep the water level consistent.

4   To maintain hygiene, bees will relieve themselves on the wing as they leave the hive, so avoid placing bees where they will fly over a clothesline or car. The yellow dots of bee excreta will not be welcome on car duco or drying sheets, unless of course the car or sheets are yellow polka-dotted.

5   The right amount of sun plays an important part in keeping the girls healthy. Look for a site that receives winter sun but is

sheltered from the hot summer sun. Ideally, the hive will get a few hours' morning sun in summer, as well as a few hours in winter. This will help keep the hive dry—bees, like humans, breathe out moisture. Making honey also requires dehydrating the nectar so too much moisture can be a problem.

6   Look for higher ground when choosing a hive location—steer clear of hollows, especially if fog is common.

7   Avoid areas of strong wind, especially if the wind blows directly into the hive entrance, which makes it very hard for the girls to maintain their temperature. Strong winds also make it much harder to fly and so increase the energy cost of foraging, which will reduce your hive yield.

8   Make sure the hive is accessible. There is a saying that beekeepers have bad backs and in my experience there is a lot of truth in it. Honey is heavy and you could be hauling lots of boxes around so decent access is important to prevent injury. Plan your hive location with that in mind and consider using smaller-sized honey supers to make handling them easier on your back. To allow easy access, leave about the size of a standard pallet around your hive for easy handling, and remember that each full-depth frame of honey weighs about 3.5 kilograms (7 lb 14 oz), so a beehive full of honey can easily weigh in excess of 100 kilograms (220 lb 8 oz).

It sounds like there is an endless list of things to take into account but really it just requires a little planning and lateral thinking. I often place my hives on rooftops, hidden away from prying eyes. Plus, the bees are going to be flying over people's heads so it ticks lots of boxes. Just make sure that they have some shelter from the summer sun and be sure to spread the load so that they do not damage the roof. Second-hand pallets are great for this.

Dealing with your neighbours

Neighbours can be the biggest challenge for an urban beekeeper. They've probably all seen the cartoon representation of bees swarming and chasing hapless humans and may be wary of their new neighbours. You really only have two options when it comes to negotiating with the people next door.

54

Option A

Don't tell them a thing until six months after you've introduced your bees, and then present them with some honey and tell them where it came from. When they express alarm about all the bees that are suddenly attacking them, explain they've been there for six months already. This strategy has been successful for me. I do remember hearing a new neighbour exclaim to her housemate 'Are those bees?!' but she never said anything to me. As the saying goes, ignorance is bliss.

Option B

Knock on the door and explain you plan to move in a hive. Reassure them that the bees will not bother them, explain the vital role they play in crop pollination, tell them how they are endangered etc etc . . . That is, spend hours and hours explaining why we need bees, only to walk away with your neighbours' disapproval ringing in your ears.

I know which approach I'd take.

Handling complaints

Keep bees and sooner or later somebody will complain about your horrible stinging invading insects.

Complaints are best handled by rational conversation. If this is impossible, try to get your bee club involved as an impartial mediator and ask the complainant to fully explain what the problem is. Sometimes the problem is not your bees but a perceived threat—I've had complaints about bees that were actually wasps, and also concern that a random passer-by with a bee allergy could be stung.

Listen carefully and assess whether it's a valid complaint, then work out a solution. If it's not a valid complaint or you can't reach an agreement or come to a compromise, your options are whittled down. The person could complain to the council or some other authority, and from there the results can be unpredictable. Some authorities will try to resolve the problem; others just want to be rid of you and your pesky bees. You can offer a jar of your finest local honey to all and sundry but sometimes nothing will make them see sense and you are going to have to move your bees.

George & Charis

These two veterans have been keeping bees for more than 30 years in inner eastern Sydney. Originally from Switzerland, George is an avid motorbike enthusiast and they rode their BMW motorbike from Spain to Australia, where George literally woke up one day and decided to keep bees.

George makes the most amazing mead from his honey and jealously guards his recipe. He can often be seen atop his terrace house roof, dressed up in his homemade bee gear and tending his hives with Charis by his side.

5

EQUIPMENT AND HOW TO PREPARE IT

Now that you've been to your bee shop and picked up all the gear, you'll be raring to go. There are plenty of online videos that show you how to assemble beekeeping woodware and they are a good place to start. In this chapter I am going to cover assembly of Langstroth gear, which is the most common hive type used by beekeepers, but the principles are the same for pretty much all gear.

Types of hives

In the wild, honey bees can live in tree hollows, hollow walls in buildings, inside roofs, under cliffs, inside barbecue lids and any other protected place. Bees have even been found living in the middle of hedges, in building eaves and tree branches, which sometimes snap if they can't take the weight.

The first man-made beehives were called skeps and were basically a wicker or straw basket, or sometimes a terracotta dome. Inside these structures everything was mixed together by the worker bees: brood, honey and lots of 'brace comb', the comb bees build in odd places, joining it together, which is why the hive often had to be completely destroyed to extract the honey.

Previous page: The J-style hive tool is an excellent choice as it can be used as a scraper and lever. Opposite: Basic beekeeping equipment: a smoker, a brush and a J-style hive tool.

Bees' usual behaviour is to glue up any gaps within the hive, except the exit/entry gap at the bottom, using brace comb and propolis. Nobody knows why. But when bee space was discovered in the nineteenth century, beekeeping was revolutionised. If a hive has internal gaps that are around 6–8 millimetres (¼–⅜ inch) wide or deep, the bees don't glue the space up and the hive cover and internal components can be removed easily.

While there are many types of hives in use across the world there are three common ones that are used by beekeepers. The most common is the Langstroth hive. You may also come across top bar (also known as Kenyan) hives and Warre hives.

A top bar hive is a great option for a backyard beekeeper, especially if they have limited mobility or strength. The top bar hive is on a stand and waist high, eliminating the need for lots of bending and lifting. Instead of frames for the bees to build their comb, top bar hives have a narrow strip of timber with a wax starter strip, or sometimes just a bevelled edge.

The Warre hive's design originally had no frames and included a peaked roof and a blanket box at the top, which was full of wood shavings designed to help maintain heat and humidity. In countries such as Australia with warmer winters, these features are now often eliminated and beekeepers use a flat roof instead. Modern Warre hives now often have frames (but with no foundation, the wax sheet that guides bees to build their comb). Unlike a Langstroth hive where the super is placed at the top of the hive, in a Warre hive you place it at the bottom. The bees follow their pattern to put honey at the top and brood at the bottom and slowly move into the bottom box, meaning you can take honey from the top of the hive. The shape of the hive has also been adapted slightly for use in Australia and consists of a square rather than rectangular box.

Hive components

The modern box hive, whatever its style, consists of several parts.

1   The hive rests on the three rails of the bottom board (risers). These rails are usually ⅜ inch or 1 inch (9.5 millimetres or 24 millimetres) thick. The open side is the hive entrance. This gap can be reduced to allow the bees to more easily defend their hive.

Opposite (clockwise from top left): Hives should be kept on a slight tilt to assist drainage; a skep hive, wild beehive on a tree, Warre hive located in a chicken pen to help keep small hive beetles under control.

2   The standard hive body, where the bees conduct all of their day-to-day nursery activities, is a full-depth box called the brood chamber or deep. This is where the queen and her ever-increasing brood live. It holds either 8 or 10 frames of comb and if it becomes too crowded, the bees are more likely to swarm. If the colony seems to be full of brood with all frames 'laid out', frames laid with eggs and containing developing brood, beekeepers provide extra hive bodies for the brood chamber (see page 95; Adding a super, for more information).

3   A queen excluder is sometimes placed above the brood chamber to keep the queen inside and exclude her from the rest of the hive—that way she can't lay eggs in the honey supers. Slots in the excluder are wide enough for workers to go back and forth but too narrow for the queen and drones to pass through. Some beekeepers sarcastically call them honey excluders because of the extra effort required by the workers to pass through them—if the workers don't go into the super, there will be no honey stored in it. It's up to you if you use a queen excluder. I find it gives me peace of mind that the queen is in a known part of the hive and can't be inadvertently damaged, killed or lost. Excluders can be made of bamboo, metal or plastic, but make sure you choose one with nice smooth edges that won't damage the bees' wings when they squeeze through.

4   Boxes above the brood chamber are called supers. This is where the honey is stored. They come in a choice of depths and widths. Depth-wise, you can choose from a deep super or a multitude of smaller, shallower sizes. You can also choose between an 8- or 10-frame wide super.

A deep, 10-frame wide honey super is a heavy beast and many beekeepers opt for smaller supers to keep the weight down. If you are using the more shallow, lighter supers, choose half size as it's useful to stack one on top of another as the two supers are exactly the same size as one deep super and can hold a single full-depth frame. This allows you to lift brood frames above the excluder if they are old and you intend to replace them (described on page 83; Swapping out old frames), a task that should be done every year to keep the brood frames new. Shallow supers are also convenient for harvesting small honey yields because you don't have to wait as long for the frames to fill before they can be harvested.

Opposite (clockwise from top left): Ready to assemble the lid and base; frame and wax foundation; queen excluder; and waxed frame and foundation.

5   Every hive has a lid—a waterproof cover. If the cover rests on the edges of the top super, it's called a migratory cover. If it slides over the top and extends beyond the top super with overhanging sides, it's called a telescopic cover. Usually with a migratory cover, a mat of linoleum or ply is placed on top of the frames under the cover. This slows the bees' build-up of comb under the lid. If it's a telescopic lid then you must have an inner cover that performs a similar duty and allows easy removal of the lid.

Frames and foundation

Frames are the internal verticals that hold the comb. They are engineered to precise dimensions that follow bee space and comprise a top, a bottom and two end bars glued and nailed together. They help keep comb building regular and allow easy inspection and honey removal. All frames are the same length, but there are different depths to suit the size of super you are using. Frames can be used with or without 'foundation', a wax sheet rolled out thin with an embossed honeycomb pattern on it. Frames often have wires stretched from side to side to help support the wax.

In Australia the wax is very clean due to little or no use of chemicals in beehives—there is no varroa mite, and insecticides are not used and so don't build up in the wax—so as long as it's Australian foundation, it will be clean. In other countries, the use of chemicals in beehives is common and the wax can have high loads of chemicals; many beekeepers avoid foundation because of this. You can get your own wax milled into foundation if you wish.

Frames get a lot of stress so make sure yours are well made. Let me say from the outset that most people find frames a pain to assemble. It's definitely easier to purchase ready-made equipment if you're not at all handy, and frames can be purchased assembled and wired. If you opt to do it yourself, note that wiring is a tricky procedure and you may need to get assistance from your club on the finer points; the wires should be guitar tight.

67

Opposite: Waxed frames and top bar with starter strip. Langstroth frames can also be used with a starter strip and no wax foundation.

Assembling woodware

If you've decided to assemble your own equipment and now have a pile of woodware in front of you *sans* instructions, the first thing to do is to separate all the pieces into piles. You should have a number of pieces of timber with handholds in the sides—these make the supers. Next, identify two large, flat pieces of timber—these are the bottom and the lid.

Your risers are the three or four pieces of timber either ⅜ inch (9.5 millimetres) or 1 inch (24 millimetres) high. These go onto the bottom board to let the bees walk under the frames. You may also have two pieces of timber or plastic that are 35 × 60 millimetres (1¼ × 2½ inches) or similar. These go under the bottom board to keep it off the ground.

Some people use wood preservative on the boxes and their joints. I do not as I am concerned about it finding its way into the honey. Make sure you use polyurethane glue for the boxes, not just a PVA glue, as polyurethane glue expands on hardening and will seal gaps better. You don't need to use glue but if you opt for no glue I would be sealing all the joins with wax prior to using the box. You will need a set square, and I use a strap clamp to keep it all together during assembly. Before you start, just make sure the frame rest groove is at the top on both ends and that all the handholds are on the outside . . . a common error.

It helps to assemble everything on a flat surface; it needs to be square in all directions before you start nailing or screwing it together. I use stainless steel screw shank nails (decking nails), which won't rust or pull free, and a nail gun to drive in all the nails. Once the glue has set I paint the outside of the box with three coats of water-based paint. Linseed oil can also be used but the downside is that it often remains sticky.

The boxes can be wax-dipped as an alternative to painting, something I don't do myself as I am concerned with the fire risks involved in mixing up a huge vat of molten hydrocarbon that's so hot it's almost at flashpoint, and doing this in my inner-city home.

The base and lid can be nailed or screwed together next; there are no real tricks to this, just make sure everything is square otherwise you'll end up with gaps. I usually cut one of the timber feet down slightly so that the hive has a permanent lean forward. You don't need much—I suggest about 5 millimetres (¼ inch). If

Opposite: Strap clamp and set square—getting tight and square before nailing.

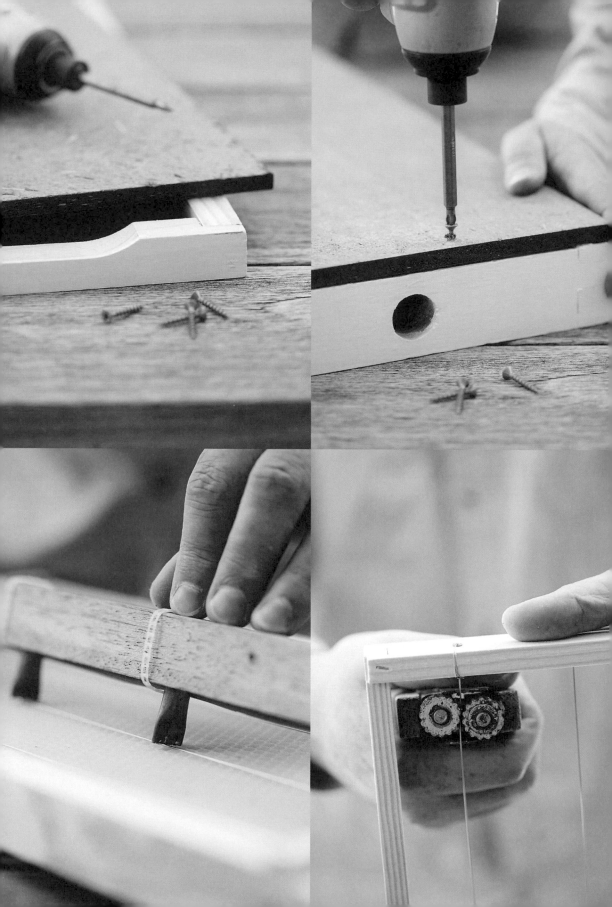

your lid comes with a tin over-lid you can put sheets of insulating material underneath (I use sheets of newspaper) to keep some of the heat out of the hive in summer and to retain heat in the hive in winter.

It's best to use a jig to assemble the frames as they must be very square and very strong. A jig helps with wiring the frames taut, too. Ask your local bee club if they can assist with a jig. When assembling frames I use glue and nails, making sure to nail the sides of each frame to ensure a strong assembly. Honey bees in Australia make a lot of propolis, which makes it very hard to remove badly constructed frames—they can easily separate on removal, making a mess of broken comb and honey in the hive.

Frames often come with sheets of 'foundation', rolled beeswax sheets with a honeycomb embossed on both sides, which gives the comb extra strength, reducing the stresses of honey extraction in a mechanical extractor. Foundation can easily be installed with an electric embedder, available from your local beekeeping shop or hand-made—there are designs online.

Insert the foundation sheet into the groove in the top bar and turn the frame so the foundation is supported on an embedding block (a block of wood that fits under the frame and supports the wax), then apply a 12-volt current to each wire; the current heats up the wire and melts the wax around it, sealing the wire in place as the wax cools again. There are many other ways of achieving the same result but this is the quickest and easiest.

An alternative to foundation is to install a starter strip in the frame: this is a 1-centimetre (½-inch) wide strip of wax melted into the groove at the top. The bees can then build their own comb from here, and it can be quite successful if interspersed with fully drawn frames—frames on which bees have created the honeycomb structure—that the girls can use as a guide to keep them straight.

Note: if you are producing cut comb—honey-filled comb sold in square blocks to eat—it's best not to use foundation because the result will be too thick.

So there you have it: you have assembled all your gear and are ready to go. Or, if you are short on time or hopeless with a hammer, you can buy fully assembled gear, which is much easier and frankly, if you only have a couple of hives, it's the way to go.

Opposite (clockwise from top left): Assembling the base; assembling the lid; crimping frame wires to get them guitar tight; embedding the wax with a home-made electric embedder.

I have only covered the basic equipment here; beekeepers have a reputation for invention and have dreamed up all sorts of different gear over the years in every shape, colour and material. Plastic is often used nowadays for frames, foundation and even the hives themselves, but the bees don't seem to like plastic and I am not a big fan either. Plastic hive bodies tend to slip off each other, which is a most unwanted feature when moving beehives, so I recommend timber rather than plastic unless you like running screaming around the yard at night while being stung in all manner of places.

Lighting the smoker

One last thing you need to know: how to light your smoker.

This simple act befuddles many beekeepers. There is the simple way and the cheat's way.

The simple way involves getting your smoker fuel (pine needles, chemical-free cardboard, chemical-free hessian sacks) and a box of matches. Take a handful of your fuel and light it with a match. Wait until it's really burning and then drop it into your smoker. Keep puffing the bellows and slowly add fuel until you have a good fire going with a nice layer of embers, then stuff in a heap more fuel while still pumping the bellows and pack tight. You should have a good, cold smoke issuing from the smoker now and you can put the lid back on. Pump it every now and again to keep the fire going; just make sure it's only cold smoke coming out with no sparks, as the bees don't like hot smoke.

Don't forget, the smoker is just used to mask the bees' communication and to suggest there is a fire nearby so they are preoccupied. You are not trying to smoke bees out of the hive, and too much smoke will aggravate the bees.

The cheat's way (which is my way) is the same but I use a small propane torch to light the fuel. You can even use your torch to light a smoker that's gone out by pushing it into the fire box and igniting the fuel at the bottom.

A good tip is to leave your smoker lying on its back on the bellows: the smoke seems to last longer.

73

Opposite: Lighting the smoker with a propane torch is quick and easy. Pine needles are a great fuel. Once you have a good fire going, the smoker should be stuffed tight and full of fuel to keep a nice, cool smoke flowing.

Bruce & Rahni

Bruce is one of the founders of GoGet, a collaborative carshare business that is helping the environment by reducing the number of cars on Sydney roads.

   Not happy with just reducing the number of cars on our streets, Bruce also harbours a keen interest in honey and bees. He built a compact garden oasis in Newtown in Sydney's inner west and contacted his local beekeeping shop. He had hives delivered to go on his roof, then took a beekeeping course. His first harvest of honey was 32 kilograms (70 pounds) and he couldn't have been happier.

6

BEEKEEPING SEASONS: SPRING

F ollowing nature's lead, beekeeping is extremely seasonal and the next chapters chart the seasonal tasks you will undertake as a beekeeper. Bees are fascinating little critters and I urge you to grab your favourite beverage and sit near the hive watching the comings and goings. You will learn as much from the outside of the hive as the inside.

It's a good idea to get yourself a logbook or hive diary so you can log all your observations as you inspect your hive every three weeks, or more often if the nectar flow is strong. This will help remind you what was going on the last time you looked in your hive. Your log will become invaluable as your hive numbers grow—beekeeping is addictive and you always end up with more than one hive.

Your log should include the date of your inspection and everything you saw. A typical log might state:

15 February, 2013, 10 am. Hive was very busy, lots of drones, saw the queen. About 5 frames of capped honey. All foundation drawn, bees were very calm.

Remember to read what you've recently logged. It helps to read the last entry before you do your next inspection, so you know for instance if you should be looking for queen cells

Previous page: A foraging bee. There is an amazing array of pollen colours inside a hive.

that may indicate swarming. I know a beginner beekeeper who repeatedly logged that he saw queen cells and then was surprised when his hive swarmed. He should have read his logs.

I also log the source of my hives and queens, which I find useful when trying to establish the genetic lines of my bees.

Hive health inspection

One of nature's expert pollinators, bees know that when spring rolls around flowers will be appearing, so they start building their numbers to collect the all-important pollen and nectar, the staple foods of the hive.

Your first job for a spring inspection is to check that the hive came through the winter period intact and is ready to support the increased numbers. So dust off your smoker and have a look at what the girls have been up to over winter.

It's probably been a while since your last check, so there may be a bit of propolis and 'burr comb', the irregularly shaped bridging comb, to contend with. Give the entrance of the hive a couple of puffs of smoke and then wait for 30 seconds, enough time for the smoke to mask the bees' communication. Crack the lid and repeat puffing in the gap, then close the lid for another 30 seconds. You will probably need to give a few puffs of smoke now and then as you go through your inspection. Smoke is useful to get bees to move off hive components so you don't squash them when reassembling the hive.

Remove the lid and check it for moisture. There may well be a bit of it so empty it onto the ground. Place the lid upside down beside the hive so that you can place supers on top of it as you inspect them.

Next, remove the mat or inner cover, if in use, and check for hive beetles, killing any you catch with your hive tool or fingernail. Look between the frames. What can you see? Bees? Burr comb?

Next, loosen the second frame on the side closest to you—use your hive tool to push the frames apart. I start with the second frame as the side frames are often glued to the sides of the box. Removing the second frame first allows you to then remove the side frame without tearing it. Using your hive tool cut out any bridges the girls might have built that could tear the frame or damage the brood, and carefully lift it. How much honey is there?

Check all the frames in the super. Try shaking an uncapped honey frame horizontally down over the hive. Does nectar fall out? If it does, then the bees are bringing in new nectar and making new honey.

Count the number of capped honey frames. If there are more than about four deep frames worth with capped honey and there are signs of honey coming in, you can harvest the other frames to give the girls some space to store their spring honey. If you live in a cold climate where the temperature has not risen much yet, you might like to leave the honey until the risk of a cold snap or false spring has passed—you don't want to remove the bees' food stores and have them starve because nothing is flowering.

The first super you remove should be placed on top of the upturned hive lid to keep it clean. As you remove each super, place it on top of the last one but rotated slightly, which helps when you are putting the supers back; they can stick together and by angling them you can more easily put your hive tool or toe between supers and the lid, picking up one box at a time.

Next, proceed to the brood box. Lift the queen excluder, turn it over and check for the queen bee as you don't want to accidently dump her on the ground—she might get lost or worse, you could transfer her to your honey supers. If she is there, transfer her to a cage while you complete the inspection or help her to walk into the brood box so you know where she is. I find that a queen is easy to spot by the way she moves, which is quite different to normal bee movement: she usually runs much more quickly than other bees in the hive, and more erratically. If she's been marked then it's even easier to find her. Queen marking, where a small dot of paint is put on her back, is used to make queens easier to locate in the hive. There is even a colour code for the dots so you can tell her age but many beekeepers just use white.

Next, place the excluder upside down over the last honey super you removed. I mark the top of the excluder so I always know what side to put down when replacing it. You would be amazed how many times I have seen the queen on the excluder, or walking around outside the hive after falling off it during an inspection.

Start with the second frame, and again cut any burr comb to make sure you don't damage any brood. If you didn't see the queen on the excluder then she could be anywhere in the brood box, so be very gentle lifting these frames in case she is on one of them.

80

Opposite (from top): Larvae and capped brood—if you can see eggs there is no need to find the queen; burr comb on top of frames needs to be removed before you lift a frame so you don't rip the honeycomb open on the way out.

When you can see an arc of honey, some pollen and then capped brood, grubs and eggs, you don't need to see the queen to know she is there; if you can see eggs, she was there no more than three days ago.

Sometimes eggs are hard to see as they are tiny. If your eyesight's not great, you might need a magnifier—the flip-down ones are good. You also need to get the light just right shining down into the cells.

Place the frame on its side beside the hive or in a frame holder and then proceed to the next one; moving one frame out allows you better access to the bees without the risk of rolling or squashing a bee.

While you are there, look for signs of bee diseases such as chalkbrood, European foulbrood and American foulbrood. Check the diseases chapter on page 141 for details and next actions—you need to do this sort of inspection at least once per quarter.

For all of spring and into the beginning of summer you are also looking for signs of swarming; the most obvious indication is a queen cell. You will usually find queen, or swarm, cells at the bottom of a frame and often there is more than one, sometimes many. Don't panic, just read the section on swarming on page 85. If you find part-made swarm cells with no egg in them, they are just being made for later and are of no concern. It's the fully constructed swarm cell with egg or larvae inside or capped that you are looking for; sometimes they are hiding under lots of bees so look carefully at the bottom of every brood frame.

Swapping out old frames

It's a good idea to change over two brood frames a year with fresh frames, and spring is a good time to consider doing it. It's necessary to replace old brood frames because every successive bee birth leaves behind a little silk cocoon that is subsequently coated with wax, ready for a new egg. Over time the cells get smaller and smaller, along with the bees born in the cells. You can often tell how much care has been taken by a beekeeper by the size of the bees: if they haven't swapped out any brood frames over the years, the bees are tiny.

Another reason for swapping out old frames is to keep the wax fresh. Disease spores and other unwanted elements can build

Opposite: Inspecting a Langstroth hive with a J-style hive tool. This style of tool gives excellent leverage to remove propolised frames.

up in wax over time, so it's a good idea to change them over. Also, if you find dark wax when extracting honey, be sure to put those frames aside—they're old and need to be replaced to help keep the hive healthy. You can recover the wax by melting the frames down with a wax melter (see page 194 for more on wax melters).

If you are taking frames from the brood box, consider taking the oldest, wonky frames but don't take frames with brood. At this time of year take empty frames or perhaps the wall frames and move them up into the honey super so they will be filled with honey and can be replaced after harvesting.

If you have seen eggs or the queen and there is no sign of disease, reverse the procedure and put everything back as you found it. Don't forget to turn the queen excluder the right way up.

Phew, there you have it: your first spring inspection done. Congratulations on your first bee wrangling adventure.

Starvation and feeding

Spring is a time of great risk for bees. The danger is of a false spring or a very wet spring, where the hive has built its numbers but is unable to forage because either there is nothing to forage on or the weather prevents it. So consider carefully how much honey a hive has in spring and whether to take or leave honey. If in doubt, leave it for them: it won't go off and it gives the hive, and you, an insurance policy. As a rule, I leave four deep frames (or the equivalent in partial frames) of honey for my bees because the winters are mild where I live. If you live in a cooler climate where the temperature drops in winter, you may need to leave a whole box. The frames of honey are not left as winter food but as food for the spring growth, when hive numbers grow but local flora might not be in full flower yet.

If there is very little honey and no fresh nectar there is a real risk of all the new bees overtaking the food supply and the hive starving. This is a sad sight: rows of bees with their heads down in the cells, dead or dying while looking for food—a definite guilt trip for the beekeeper. Head this off by feeding them.

Feeding is easily accomplished using either a commercial feeder or a large (at least 1 litre/35 fl oz) zip-lock bag filled with feeding mixture placed on top of the hive. See page 133 for details on feeding and the mixture to use.

Swarming

What a word: swarm.

If I had a bee for every time I have heard that the bees are swarming on something, I would have my own bee swarm. In Greek mythology, a bee swarm landing somewhere conveyed a blessing. Somehow we have forgotten this and relegated it to a curse. A bee swarm according to the general public and some authorities is to be feared and exterminated at any cost.

Time for some rational thinking. Bees swarming are usually at their most passive and least likely to sting because they have gorged themselves on honey and are so full they can't bend to sting you, plus they don't have a hive to defend.

So what's a bee swarm and why does it occur?

Bees are geared to replicate themselves, so when they feel the conditions are right—there's plentiful food, a warm ambient temperature and a big bee population—they prepare queen cells in order to replicate the hive. The swarm when it finally forms will consist of the old queen and about half the bees, who will have gorged themselves on honey. Normally queens cannot fly, so the bees starve her for a few days so she slims down.

The queen cells start much like any other cell and are usually found at the bottom of frames. As they grow they start to resemble an unshelled peanut. The workers then either encourage the queen to lay in these queen cells or move an egg from another cell into one of these. A queen is the quickest bee to develop, taking approximately 14 days from egg to hatching, and moving an egg into the queen cell sets the clock ticking, determining when a swarm will be produced. Once the bees have decided to swarm little will stop them. You can destroy the queen cells as much as you want but they will just make new ones. It's also possible that you won't spot all the queen cells: I have seen 12 cells in a single hive and still missed one hidden in a frame corner. Once you find queen cells you really have two choices: let the hive swarm or perform a split.

A hive split

When splitting a hive you need to check every frame for queen cells and split the brood so that half stay with queen cells in the old hive and half are put into a new hive, along with the queen. You can place your split hive beside the other one. The girls will sort it all out and within a couple of weeks you should have a new queen laying eggs and two prosperous hives.

Surround the brood frames in both hives with a frame of honey on either side, followed by either 'undrawn' frames or empty 'drawn' frames (frames where the bees have constructed honeycomb). Make sure that each hive has approximately half the bees of the original colony and include in each a variety of bees of all ages. If you have a lot of queen cells you might like to pinch a few off to stave off the possibility of a swarm if the new queen doesn't kill the other queens when she hatches. Make sure you leave one or two of the most advanced queen cells and well formed—you can tell they're close to hatching because the tip of the cell changes colour to a golden brown.

The hive with the queen cells will behave differently for a while; a queenless hive has a different sound, a different buzz. If you look into the brood frames a queenless hive often stores new honey in the brood cells but as soon as the queen hatches, the workers will clean these cells out ready for eggs.

Don't be surprised if it takes a couple of weeks before you start to see new eggs and brood; it can take that long or even longer for the queen to hatch, develop, mate and start laying. Don't panic if after four weeks there is no sign of brood forming in the hive that had queen cells—check if they have hatched. If they have and there is no sign of a queen you can join the two halves back again.

A good trick to check for a queenless hive is to take a frame of brood with eggs from another healthy hive and put it in the queenless one. If they start making queen cells there is no queen; if they don't make queen cells then there is a queen. If it's definitely queenless, you can either purchase a queen or merge it back with the old hive.

To merge it back, open the hive with the queen and place a couple of sheets of newspaper or similar over the hive and cut a couple of slits in the paper with your hive tool so the odours permeate. Then, place all the boxes from your queenless hive on top of the newspaper, starting with the brood box and followed by the honey supers. Put the queen excluder to one side—you only need one and as your hive is queenless a second one is of no use.

Put the lid on top of your now possibly really tall hive (hopefully you won't need a cherry picker to do this . . . don't worry, you'll be able to remove the extra boxes once the honey in them is capped and harvested). After a few days you should see bits of paper being ejected at the front of the hive: the hives have now merged back into one united community with minimal, if any, fighting. And you can let the bees fill the old brood box with honey to be extracted.

If you have purchased a new queen, my favourite method of introducing her into a queenless hive is to first make sure the queen cage has had the closure (tape, plastic plug or cork) over the candy (a mixture of fine sugar and honey in the entrance of the cage) removed. Then, place the queen cage in between a couple of brood frames, making sure the exit is tilted up so that any dead escorts don't block the exit. I give the top of the brood frames a light spray from a spray bottle that has a few drops of vanilla essence in it to mask the new queen's pheromones. Leave the hive alone for seven days, during which the girls should release the queen and all should start ticking over again. Sometimes the new queen is not accepted and will be killed by the worker bees. Don't be alarmed; you can always merge this hive with another using sheets of newspaper as outlined above.

Elke

When Elke decided to take up beekeeping, it took a couple of attempts before bees would accept her. We tried a swarm first, and then a feral hive that took hours in the sun to cut out of a tree, but no luck—the bees would not stay.

Bees have finally accepted her and she now has a couple of hives perched in her garden paradise in eastern Sydney. Elke is an active member of the Sydney branch of the bee club and a horticulturist with a keen interest in plants and trees. Being a beekeeper, this is good knowledge to have up your sleeve. Her work as an arborist has her looking at trees regularly, and she can now keep an eye out for bees as well.

7

BEEKEEPING SEASONS: SUMMER

Wow, it's summer! Everything's blooming, swarming has pretty much come and gone, and the bees are in full production mode with big populations bringing in a good nectar flow.

Hive health inspection

Check your logbook to see what the girls were up to at the last inspection. Once summer has arrived you should already have performed your brood inspections and be onto monitoring the space for honey storage, and perhaps lifting a brood frame every alternate inspection to make sure you have healthy brood.

So let's get kitted up. Light the smoker and get cracking with perhaps the most exciting hive inspection—the money shot of backyard beekeeping (or perhaps it's the honey shot).

Always kit yourself up fully, don't get complacent or brave—the bees are really good at knocking you back down when they think you're getting a bit too smart for your own good. A carefully placed sting to an eyelid can teach you all sorts of respect.

After a couple of puffs of smoke in the hive entrance and under the lid as per usual, open the lid to the hive. What do you see?

Previous page: Beekeepers about to inspect a hive. Opposite: Burr comb in the lid of a hive and on top of mat, sometimes a sign of limited honey storage space.

Have the girls filled the lid with honey and burr comb? If there is honey in the lid and you use a hive mat, it can be taken as a sign that room is tight and the girls need a super. Take the lid off and place it on the ground beside the hive, top down.

Cleaning out a lid filled with burr comb is a pain as it's always got bees all over it and they are not keen to leave. An escape board (explained on page 98; Removing bees from frames) placed over the lid is a great way of emptying out the bees while you scrape out the comb into a bucket.

If you already have a honey super in place, look down between the frames. Have all the frames been drawn? Can you see honey at the top of the frames? This is a good indication of space and flow: if all frames have been drawn, then you know the girls are looking for more storage space. Next, lift the frame second closest to the side of the box. What do you see? Fully capped honey or only partly capped?

Honey can't be harvested from a hive until the frames are 70–80 per cent capped, so go through the frames and count them. If all are full but not capped, it's time to 'undersuper', or add a super (explained below) to give the hive more space. If only half of your frames are full, there is time and you can undersuper at the next inspection. Of course if your hives are under some trees that produce a lot of nectar, things can happen very quickly and you might need more room sooner. Only experience in your area can tell you what the nectar flow is like.

95

Adding a super

To 'undersuper', you simply take a super of undrawn or drawn frames and place it under the partially filled honey super, not on top. Bees will store honey at the top and work down so if you place your empty super on top the girls are likely to abandon filling your partially full super and start filling the empty top one instead, a most annoying circumstance as you now have two partially filled supers and still can't take any honey.

If your hive consists of only a brood box (perhaps it was a swarm or new hive split this season), then your assessment is slightly different. If there is honey stored in the lid it's a good sign of progress and perhaps a queen excluder and honey super can be added.

Opposite: Removing burr comb from the hive lid. Wild comb like this makes an excellent gift.

If the whole brood box is in use with all frames drawn and lots of capped brood, it's definitely time to add a brood box. If your brood frames and honey frames are the same size then you can assist the bees in moving up into the honey super by removing one of the wall frames from the brood box, making sure it only contains honey, and placing this frame in the middle of the new honey super directly above the brood. Bees will be more inclined to start using the new super if they already have stores there. Replace the frame with a drawn comb, if you have one, or an undrawn frame.

If you are using a queen excluder with your hive, make sure you shake all the bees off the frame into the brood box—you don't want to move the queen into the honey super. If this happens you have really mucked up the hive: with no queen in the brood box the girls will sometimes raise some new queens but of course yours can't leave because she is above the excluder. Everybody seems to get confused and you also have the problem of brood in your honey supers.

Having assessed and added space where necessary, close the hive up and leave them to it. Hopefully you have honey on the way and will soon be harvesting.

Don't worry if there is not heaps of honey there yet; be patient, it will come. I get asked all the time when the honey will be ready and I usually shrug my shoulders and say it will be ready when the bees are.

Harvesting

Once you know you have honey to extract and will be harvesting, then comes the reward of beekeeping. It's also the messy part; some beekeepers have found honey creeps into everything they own.

You need to put a bit of thought into harvesting: select a location where you can extract the honey; gather all the necessary equipment; and allow enough time so that you're extracting the honey as soon as possible after removing frames from the hive, while the honey is nice and warm and easiest to extract.

First, prepare the extraction area in your kitchen or some other area that can be cleaned easily and is bee-proof—honey tends to get on everything you've used and the girls will smell the honey and come calling. There will be bees in the room anyway but you don't want the girls stealing the honey back when you're not looking.

You will need some food-grade buckets and a strainer of some sort (I use a mesh stainless steel colander). It's also useful to invest in a plastic tub with a lid that's big enough to hold your honey frames. I use food-grade versions of these purchased from catering supply shops as the hardware store ones are just not strong enough.

You also need an extractor or spinner, a hot or cold uncapping knife and a scratcher, all of which can be bought or hired from your local bee club. An extractor is a washing-machine-like centrifuge that literally spins the honey out of the frames. The hot or cold uncapping knife is used to cut the cap off the top of the honeycomb and let the honey out. The scratcher looks like a really sharp and wide fork and you'll use it to scratch the capping off the bits that the knife can't reach.

You should extract the honey as soon as possible after removing the frames from the hive, not only because warm honey is runny and easier to extract than once it's cooled, but because extracting soon after removing your frames also reduces the risk of the small hive beetle, which lays eggs that will hatch in four days, getting into and destroying your honey. So don't pull the frames from the hive and leave them for a week, or you might come back to a festering mess or candied honey, which is very difficult to remove.

Now that you've got everything you need, you're ready to harvest the honey. Open the hive in the usual manner by giving the entrance and under the lid a couple of puffs of smoke. Place the lid top down next to the hive and inspect each honey frame, looking for frames that are three-quarters capped. You can also try holding frames on their sides with the cells open-ends down and giving the frames a good shake. If the honey is ripe it should stay in the uncapped cells. Don't take any frames from which nectar falls out: it's not ready to be taken and will probably ferment.

Once you know how many frames of honey you can take, work out how much you are leaving. The amount you leave will vary depending on your local conditions and the strength of your hive, so talk to your local bee club members. In summer I leave at least the equivalent of four full frames, spread across a number of partially completed frames, just in case the weather turns bad for an extended period; you are taking the bees' food after all and they must be left with enough or they will starve.

Removing bees from frames

But how do you get the bees off the frames? There are a number of ways you can do this. I use the shake and brush method but you could also use an escape board or a bee blower (not for the faint-hearted).

Shake and brush method

A bee brush is a brush made specifically for removing bees from frames. Don't be tempted to grab any old brush from your shed—remember you are dealing with food.

To shake and brush, select the frame you wish to take and, holding it in both hands by the sides (called ears), give it a couple of hard shakes over the entrance of the hive or over the open hive. Most of the bees will be dislodged into the hive; the remainder can be brushed off with your bee brush. To brush bees, use a slightly wet brush and brush the frame upwards so that any girls that are head down in a cell get brushed up and out of the cell rather than against the edge of a cell. Keep a bucket of clean water handy to keep the brush clean as it can become covered in honey and not be as efficient in removing the bees. Place the bee-free frame into your tub and proceed onto the next frame until all the frames have been taken. The shake will take some mastering but it is a very efficient way of removing bees from a frame. If you empty the entire super, remove the super and place it beside the hive until you return the empty frames (called stickies).

This method is a great way of removing a super of frames. The downside is it can be slow if you have many hives to harvest. And because the hive is open for a while the bees can get annoyed, and nobody likes annoyed bees. Robbing, where an adjacent strong hive starts attacking another hive and stealing the honey, may also start; if that happens, close the hive and give up for the day. Things can get out of hand very quickly and bees get quite aggressive during robbing. The only solution is to close up the hive and reduce the entrance of the hive being robbed to a narrow opening so it's easier for them to defend, and leave them to it.

Opposite: a frame of honey almost ready to harvest. A good shake in front of the hive will remove most of the bees.

Use grass or leaves to reduce the entrance because the girls can chew it and remove it themselves; you don't need to do it for them.

Escape boards

An escape board is a flat wooden board placed between supers that has a one-way valve in it called a bee escape. This can take many forms but essentially it allows the bees to get out of the super and not return. Using escape boards is a great way of harvesting frames without any agitation to the bees as the supers will be mostly bee-free. They work best when it's cooler and the honey is capped; if the super is still being ripened they are not as effective.

The actual escape design has several variations from a rhombus maze to a porter butterfly and many more besides. Try a few and ask some other beekeepers: they will all no doubt have an opinion, as beekeepers do!

To use an escape board, insert it between the super to be removed and the partially empty super below. Do not use an escape board unless there is a partially empty super below your full one, otherwise the bees will have nowhere to go and the hive could become overcrowded. After about four hours most of the bees will have left the super and you can shake and brush the remaining bees. After 24 hours even more bees will have left but of course with no bees to keep them under control, the small hive beetle will be having an egg-laying free-for-all, so don't wait too long.

I don't use escape boards because it requires two trips to the apiary to remove a super of frames, and I am also concerned about the small hive beetle eggs being laid in the supers. However, they are useful if your hives are in your backyard and close to neighbours: there is minimal disturbance to the bees and it's less likely that they will become agitated. Escape boards can also be useful to allow bees to leave a super of frames that you have previously removed from a hive.

Bee blowers

A bee blower is essentially a leaf blower and is really only for the experienced beekeeper—I am mentioning them here so you know what people are talking about. I do not think they are a good idea unless you know what you're doing, and even then it cannot be good for bees to be blown about. Plus, if there are passers-by, they could get stung.

Previous page: Watching the bees should be the first step in your hive inspection before opening the hive.

Fume boards

A fume board, available from some beekeeping shops, is a board sprayed with a special solution that bees don't like; many smell a bit like marzipan. To use a fume board, put the recommended amount of solution on the special lid (the fume board) and then place the board on the hive. After a few minutes the bees come running out as they can't stand the smell and you can take the entire super, empty of bees, off the hive.

Extracting the honey using an extractor

All your frames of honey are now in your tub, so let's get the runny goodness out of those frames. Grab your tub and jog into the house . . . the first thing you realise about honey is that it's very heavy. Better get a trolley.

Our mission is to uncap the honey frames so the honey can get out. But because the clever girls make the honeycomb on an angle, it won't run out unless we spin it out or somehow use gravity to our advantage.

To get the honey out, you can either uncap and extract or use the crush and strain method (see below). If you are uncapping and extracting, take a frame and place it on its side with the top bar and bottom bar vertical, keeping hold of the topmost side bar. Position it in a pan or, if you are lucky enough to have an uncapping tray with filter, use that. If you have a heated knife, plug it in. If you don't have a heated knife, don't worry—a sharp, serrated knife works just as well if you only have a few frames to do. With a sawing motion, pass the knife down using the top and bottom bars as a guide and slowly cut off the cap covering the cells. The knife won't reach it all but a fair bit should be uncapped. Spin the frame around and run the knife down the other side. Next, take the scratcher and scratch the remaining cells open. I usually continue to uncap until I have an extractor load, putting each frame into the extractor until the extractor is full.

Now the fun part: extracting using an extractor.

Start spinning the extractor by turning the handle, or if it's electric, by pressing the button (my favourite). Not too fast—if the honey has just come off the hive the wax will be soft and the honey runny; you don't want to damage a frame needlessly. About 2.5 kilograms (5 lb 8 oz) of honey will come out of each

103

frame you spin, and before long there will be quite a lot in the bottom of the extractor.

Once one side of the frames is empty, spin them around and do the other side (some extractors such as radial extractors do not need this step). If the extractor suddenly gets hard to turn it's because a large volume of honey has accumulated in the bottom (that's a good problem), which is obstructing the rotation. Open the gate and let all that sweet goodness run out into your bucket—don't worry about straining it yet.

Alternatives to an extractor

You have a couple of options to extract the honey without an extractor.

1 Gravity

   Place all the honey frames upside down in a crate and let the honey seep out over a few days. This is a cheap, albeit slow, method. Remember, the girls constructed the honeycomb cells on a slope so it's really important that the frames are upside down otherwise the honey will be held in.

2 Crush and strain

   Crush up all the honeycomb and let the honey slowly strain out or use a fruit press to press the comb and squeeze the honey out. This method works best with warm honey and if you are not using foundation, as you can just pop your frames back in with a starter strip and the girls will draw new wax. If you are using foundation, you must re-wax your frames before using them again.

Now that you've finished extracting all your frames, you have a bucket or two of unstrained honey and a pan full of wax, called cappings, that you cut from the frames. What next?

First of all, don't throw out the cappings—they contain lots of honey that will slowly filter out and you don't want to lose that, so wrap them up in some muslin cloth and hang the cloth over a bucket; the honey will slowly drip out. You will be surprised just how much will filter out. Second, you need to strain your newly extracted honey to get the big wax flakes out. I don't filter my honey to glass clarity because I believe that there is a lot of goodness in the pollen and other matter that's left behind, so I just pass it through a wire mesh colander into my bottling bucket. The bottling bucket is simply a large bucket that has had a special

Opposite: Uncapping a frame of honey using an electric uncapping knife. You should aim to open all the cells, otherwise not all the honey will be freed.

valve called a honey gate fitted to the bottom for easy jar-filling. Bottling buckets are available from beekeeping shops and will save you lots of time as they are large, reducing the equipment you need, and with a honey gate fitted the honey flows easily.

There you have it: your first extraction. I can almost guarantee you have more honey than you expected. Hopefully you had enough buckets at the ready and didn't resort to saucepans (done before, many times).

Opposite (clockwise from top left): Finishing the uncapping with a scratcher to open any cells missed by the knife; loading the extractor with uncapped frames; spinning out the honey.

Marion & Brian

This couple owns an organic farm in New South Wales' Hunter Valley and it has allowed Marion to pursue a long-held wish to keep bees. Through contacts, they managed to find a local farmer who had some hives and installed them in a lovely clearing on their property.

Marion and Brian hope to produce their own organic honey and increase the yield from their citrus, olive and walnut trees, along with other produce like garlic and herbs. The morning after installing the beehives, the bees were hard at work, with each dandelion head bobbing as it was visited by a bee. Both native and European honey bees visit the flowers, providing a great reason not to mow the lawn.

8

BEEKEEPING SEASONS: AUTUMN

The days are starting to shorten and the temperature is cooling a little. Autumn is the time the bees set up their hive and ready themselves for winter; it's also the last opportunity for the beekeeper to have a good look at the brood box to check for disease—and the last chance for honey harvesting before winter.

Hive health inspection

Check your logbook or diary to remind yourself what the girls were up to the last time you had a look. Now let's start by doing a full inspection. Before you begin, look at the entrance. What do you see? Bees with pollen? Bees fumbling their landing because they are full of nectar? You can learn a lot about the hive by looking and listening. If the bees are bringing in pollen then they most probably have brood to support and of course bees full of nectar means there is a honey flow in progress.

On to the inspection. Give the entrance of the hive a couple of puffs of smoke and then wait for 30 seconds. Crack the lid and repeat puffing in the gap, then close the lid for another 30 seconds.

Previous page: A couple of puffs of smoke under the lid before opening the hive makes the bees move down, away from the lid.

Remove the lid and place it upside down beside the hive, ready to receive the supers as you go.

Next, remove the mat or inner cover, if in use, and check for hive beetles, killing any you can catch with your hive tool or fingernail, like you did in spring.

By now you are probably used to doing hive inspections, so take your time to observe what's going on as this information will help your future assessments.

Next, loosen the second frame on the side closest to you, cut out any bridges the girls might have built that could tear the frame and carefully lift it. How much honey is there?

Check all the frames in the super and count the number of capped honey frames.

Try shaking any uncapped honey frame cells down over the hive. Does nectar fall out? If it does, the bees are bringing in new nectar.

If there are more than about four deep frames worth of capped honey and there are signs of honey coming in, you can harvest the other frames to give the girls some space to store their remaining summer/autumn honey. If you live in a cold part of the world where the temperature really plummets to freezing point, you might like to leave a full super of honey for winter; your local bee club can tell you more about how much honey you should leave in your area.

As you remove each super, remember to place it on top of the last one on a slight angle so it's easier to access one super at a time with your hive tool when you put them back.

Next, proceed to the brood box.

Lift the queen excluder, turn it over and check for the queen bee. If she is there, either transfer her to a cage while you complete the inspection or help her walk into the brood box so you have some idea of where she is.

Start with the second frame and repeat the check for burr comb. Cut any bridges to make sure you don't damage any brood as you lift the frame. If you didn't see the queen on the excluder then she could be anywhere so be very gentle lifting these frames in case she is on one of them.

What can you see? An arc of honey, pollen and capped brood, grubs and eggs? You don't need to see the queen to know she is there; if you spot eggs, you know she was there three days ago.

Place the frame on its side beside the hive or in a frame holder and then proceed to the next one. Moving one frame out allows you better access to the bees without the risk of rolling or squashing a bee. While you are there, look for signs of bee diseases such as chalkbrood, European foulbrood and American foulbrood—check the diseases chapter on page 141 for details. You need to do this sort of inspection at least every quarter.

Preparing the hive for winter

If you have seen eggs or the queen and there are no signs of disease, it's time to think about readying the hive for winter. If you live in a cold climate you probably want to pack the hive down so the girls are keeping a small space warm, saving them energy. To pack the hive down, simply remove empty supers and leave just your honey super and brood box. At this stage I also like to move the queen excluder up under the lid as it helps with spring swarm management: it gives the bees a bigger brood box that they can expand into rather than become confined in. It also gives the bees better access to the honey stores. I then put the excluder back down after swarming has passed next year.

In cold climates, beekeepers might replace the lid with a dark coloured one to promote heat transfer. As I'm based in Sydney where the winters are mild, I don't do this. In fact I don't pack the girls down at all—I leave all the supers on but I place a board between the unused supers and the rest of the colony, with holes in the corners that allow the girls access to the upper stories while keeping warm and dry. I also lift the queen excluder so they can rearrange their house as they please. The girls do not seem to move up much during autumn and winter, which goes against what most people think but then that's beekeeping for you . . . the girls can't read and don't know what's expected of them.

Once you have readied the hive for winter, reverse the whole procedure, putting everything back as you found it. Don't forget to turn the queen excluder back the right way up.

Autumn is also a good time to make sure the hive has not been grown over by trees that will rob it of winter sun—maybe you need to do some pruning. And finally, make sure the hive has a slight forward slope to help drain any moisture that may accumulate during winter.

Opposite: Unfiltered honey from the extractor can be passed through a sieve prior to final bottling. This filters out any lumps of wax that might be in the honey.

Tessa

Honey was always a big obsession for Tessa: every morning, her breakfast of choice would be an English muffin with equally large amounts of butter and honey. After becoming interested in sustainable gardening, and becoming more aware of the impact of people on our environment, she wanted to explore ways of leaving the world a little better off.

These little seedlings of ideas came together two years ago when she moved into a new flat. There was a cat door and she really wanted a cat. However, concerned about having a cat to look after and the possibility of one cat becoming a few more, she chose bees instead and became a beekeeper, tending hives in a community garden in Lane Cove. Her second batch of honey was fantastic.

9

BEEKEEPING SEASONS: WINTER

During winter the bees slow right down, with reduced populations and reduced food needs. In some climates the bees will form a tight cluster around the brood and queen, maintaining that 35°C (95°F) temperature by vibrating their wing muscles. It requires a lot of effort to keep the hive temperature steady and the bees share the load: the tired bees move into the cluster for a rest, while the rested bees move out to be the heat producers on a constantly rotating basis.

If you get a warm day (I wait for 18°C/64°F) you might be able to perform a brief hive inspection but you really need to be quick so as not to chill the brood. Otherwise I would leave it until spring, unless you have a reason to be concerned.

Winter is a good time for beekeepers to put their feet up and enjoy some honey on toast; take a holiday from beekeeping and get plenty of rest because spring is only around the corner.

Winter feeding

During the colder part of winter it's best to the leave the girls alone and perhaps just pick up the back of the hive from time

Previous page: While inspecting the hive, placing frames to one side using a frame holder keeps them off the ground.

to time to estimate the amount of stores they have; a luggage scale can be useful for this purpose. If you measure the weight at the beginning of winter you will have an accurate measure of the stores and can plot the progress in your hive records. If you think the hive is getting too light—if it drops more than the calculated honey stores you left on—then add an entrance feeder or a feeder that can be filled without opening the hive and letting the cold air in and the hot air out. There are numerous designs and methods for doing this. Bees have been known to make it all the way through winter only to starve as spring starts because the beekeeper has seen the seasons change and thought that all would be well.

Under no circumstances should bees be fed honey—this is a really good way of spreading disease. Feed only white sugar either as syrup made with water or as fondant or even dry sugar. The best winter mixture if the daytime temperature is about 10°C (50°F) is one low in water, such as two parts sugar to one part water by weight. If you are in a colder region, consider feeding candy, granulated sugar or fondant. You can purchase baker's fondant; just make sure it contains only sugar. It's fine to shake sugar into the hive in granule form—the bees will deal with it providing they have access to water. See page 133; Feeding, for more information.

One important note about sugar feeding: it's only there for emergencies. Sugar doesn't contain the micronutrients of honey and is not a replacement. Bees should also never be fed when you are planning on harvesting as honey made with sugar is not honey, it's sugar.

Whether you need to feed your bees or not will depend on the climate. In suburban areas in Sydney the bees keep flying due to the warm winter temperatures and enjoy a variety of forage produced by the varied plantings of introduced species that flower during our winter. In fact it's possible for the bees to increase their stores rather than deplete them.

Equipment maintenance

Winter is a great time to fix up that gear that needs painting, rewire those frames that need it, make candles and otherwise take a break from your summer honey extraction commitments.

Beekeeping woodware takes a beating from the elements and every year you will have things that need replacing or repairing. During the warmer months, I usually rotate off the boxes that are worse for wear and fix them by filling holes and repainting as necessary. Remember while you are working on bee equipment that it comes into contact with food, so always keep it clean and be careful that the products you use become inert once cured.

Frames are easy to repair. First, check that the wires are tight; if they are loose you can use a clamp to compress the sides and then use a tack or two driven into the sides to loop the wire past so that when the clamp is removed the wires will be tighter. A wire crimper can also help get that twang of a tight wire. Replacing foundation wax is also quite easy. I recommend using an electric embedder—being a bit of a handyman and a bee nerd I made mine from a piece of wood, some copper pipe and a halogen light transformer. They can be bought from beekeeping supply shops and are real time savers.

The way I have maintained my equipment has changed over time. I started following advice in lots of books and treated all my gear with anti-fungal timber treatments, painted the boxes inside and out and all sorts of other time-consuming tasks. I am a lot more relaxed these days: I no longer use any sort of timber treatments and only paint the outside of the box for weather protection. My philosophy now is if I can't eat it I don't want it near my honey.

Opposite: Bees cleaning the tops of frames. You should expect to see at least this number of bees in a super during summer.

Katrina & Jonathan

Katrina and Jonathan are two keen gardeners and have a backyard taken up by large garden beds full of all sorts of food crops and fruit trees. They noticed they weren't getting good crops and put it down to not enough pollinators. After some research and attending a bee course, they installed a Warre hive in the middle of their chicken coop. This is an ideal location for a hive as chickens eat the larvae of the small hive beetle, which is a pest that can cause damage to the bee colony. Since introducing their hive, Katrina and Jonathan have had much better yields from their garden and even better local honey.

10

BEEHIVE MANAGEMENT

B ees are clever little things with an incredibly complicated society, but occasionally things go wrong in nature and when they do, a hive can die out. As a beekeeper your role is to keep an eye on things and help your bees out when they need it.

Moving your hive: it's all in the planning

Bees are remarkable navigators: they can visit a flower up to 5 kilometres (3 miles) away from their hive, go home, then go back out and find it again; they can also tell their hive mates how to find it using landmarks as well as sun location—quite an incredible feat. But for a beekeeper, this means that you can't just up and move a hive from one side of your yard to the other because all the field bees will return to the old hive location and get very confused. You and your neighbours do not want a yard full of confused bees, believe me.

So how far can you move a hive? There is a saying that it's either 1 metre or 5 kilometres and nothing in between. In practice, if you're moving a hive backwards and forwards you can move it 2 metres (6½ feet) or so and the girls will find their way home;

Previous page: Inspecting the honey super for space should be performed at least every three weeks during the warmer part of the year.

but if you're moving it side to side, move it no more than 1 metre (3 feet) every few days. If you need to move it more than that, then you'll have to first move it more than 5 kilometres (3 miles) away, so that the foragers won't be tempted to use any of their old landmarks and end up in the old hive location. And you'll need to leave it there for about 40 days before moving it to the new spot, so that any foragers who still remember the old landmarks in your area will be deceased.

So you really need to choose a good spot when locating a beehive because moving it can be a major task, especially in an urban environment where a temporary home 5 kilometres away is not always easy to find.

How do you move a hive a short distance?

It's quite easy. If it's just 1 metre (3 feet) or so, put on your protective gear, give the hive a little smoke and, together with a helper, pick it up and put it down in the new spot. You will see some bees looking for the hive, doing circles around the old location (orientation circles) but all should be fine.

What if it's a long way?

First, let me tell you a story of a hive move that went wrong.

My mate had a hive on his roof and he was giving it to me. He had locked all the bees in and all I had to do was climb a ladder to his rooftop, where he was waiting to pass the hive to me. We did the handover, but unfortunately he hadn't sealed the entrance properly and while I was balanced on top of the ladder, the bees started crawling out and all over me. They quickly found the chink in my armour—a spot in the crook of my elbow where my suit was sticking to me—and started stinging (30 stings in total). At the same time the ladder started wobbling. Somehow I climbed down and wrapped the hive up in a sheet to contain the bees, but I was still covered in hundreds of bees and they were still stinging me. My mate's wife was trying to brush the bees off me with a goose feather (not supposed to upset them), but it wasn't helping much— they were already upset!

We bundled the hive into my car and headed off to the new bee site while wearing full bee suits (this got us a few amused looks from passers-by) and all my mate could say, as we sat in a car full of bees, was 'I hope we get pulled over by the police'. It took two weeks for the swelling in my arm to subside enough to allow me to use it again.

After this experience I learned how vital it is to be organised and to do the important work, like blocking the entrance, yourself. The instructions below are based on many successful hive relocations.

The first step in moving a hive a long distance is to get your equipment together and prepare the hive for moving.

You will need the following:

1   Some gaffer or duct tape.
2   A red torch (a pushbike tail light is ideal) because bees can't see the red light.
3   A rag or some newspaper to stuff in the entrance.
4   If it's hot or humid weather, you will need a flyscreen lid to put under your usual lid. Remove your usual lid when moving the hive so the bees can ventilate; otherwise they can melt down and die from the heat they generate, a very sad thing to see.
5   A strap to hold the hive together—a ratchet strap from the hardware store is ideal.

So now you have all your gear together. On the day of the move, tape up any holes the bees might use as an exit, leaving the main entrance open; otherwise you will have a nasty experience moving it that evening. If you are using the flyscreen under the lid, place it there during the day as well. If you're not using the flyscreen, put the ratchet strap on and make sure it's tight—trust me, you don't want the hive separating during the move.

When it's dark, get your red torch and see whether all the girls are at home or if they are hanging out on the porch. If there are bees hanging around a few puffs of smoke or a very light spray of water from a spray bottle will usually send them inside. Once they are all inside, stuff the entrance with your rag or newspaper and make sure it's tight; put some tape over the stuffing as well just to be sure.

If you are using the screened lid, take the old lid off and put the ratchet strap on, being careful not to dislodge the screen or you will have bees crawling all over you like a scene from a horror movie. Make sure the strap is nice and tight. You can now put the hive in your car, truck or van. Keep it upright and point the entrance at the front or back of the car to minimise the frames rocking around.

With the hive in your car, make sure it's strapped down so it can't fall over, and move it to the new location without delay.

132

Previous page: Rooftops can be an ideal place for beehives as the bees' flight path is well above passers-by. The tree here gives shelter from the hot afternoon sun.

When it's in the new location it's a case of simply reversing the procedure. Once the entrance is unstopped, leave the girls alone for a few days before you inspect them again.

Well done—your first hive move is complete.

Feeding

Sometimes you need to feed bees. It shouldn't be part of your normal routine as the girls should have been left enough honey for winter, but in some instances—a bad season with not enough nectar, a small swarm that needs a leg-up, a sick hive—stimulating with a feed will help a weakened hive recover.

Do not feed bees honey or pollen that has not been irradiated; it could be contaminated with disease and you will be infecting your hive and possibly killing it. You could feed them back their own honey but you would need to be extra certain that it had not been contaminated with honey from another source. Over time, refeeding bees their own honey alters its composition and may have adverse affects. Your local beekeeping shop can sell you specially irradiated honey and pollen as food for your bees.

Bees need nectar for their carbohydrate and pollen as their protein so when feeding you need to make sure they have access to both. Feed them pollen by sprinkling it onto the top bars of the hive. Often the bees will eject some of this pollen out of the hive as refuse, though they will also keep some and use it. Special pollen patties are available from beekeeping shops and are designed to provide bees with all the nutrition they need.

Nectar can be replaced with a sugar solution, granular sugar or candy. Feed it to them by taking out a few frames, pouring the granular sugar or sugar solution into empty combs and placing them back into the hive, or even sprinkling the granular sugar into the hive (yep, just chuck it in), but make sure you are always feeding refined white sugar. Do not use brown or raw sugar as they can contain contaminants.

There are various feeders available for hives and each has its advantages and disadvantages. The most important consideration is that it must be impossible for the bees to drown in the feeder.

I usually use a 3-litre (6-pint) zip-lock bag with a few slits cut in its top-most side with a razor blade. I place the bag on top of the frames under the lid—you may need to add a rim or similar to give

133

you space to put the bag on top of your frames. Surface tension will keep the contents in and the bees can sip away without drowning. The other advantage is it's close to the brood so it's kept warm. Check the bag in a week and if empty, replace. The girls will stop taking this mixture when there is nectar available.

There are two mixtures to use during feeding. One mixture is used to stimulate brood rearing and closely resembles nectar. Mix one part sugar to two parts water by volume. The bees will make comb and raise more brood so it's perfect for stimulating a colony during the warmer part of the year—it's not for winter. If you want to supplement bee stores in the colder months, instead of the 1:2 mixture of sugar to water, use a mixture of equal parts sugar and water. This ratio makes the syrup easier to use as less water needs to be removed by the bees to make the syrup into a honey-like fluid.

Syrup can spoil and ferment over time—if yours starts to smell at all, discard it.

Dry sugar is easier to manage than syrup and does not spoil. Its application can be as simple as tossing it in (as long as you have a solid bottom board) or placing a container or some paper on top of the frames and piling it up. The girls will need water to use the dry sugar so make sure they have a water source. During winter water is often found under the lid in the form of condensation, which may provide them with enough fluid.

Queen problems and re-queening

There will be times when you need to re-queen. Sometimes a hive can be roaring along and then next time you look, the queen has disappeared; or maybe the queen was old. Maybe they swarmed and the new queen was eaten by a bird or got lost when mating and didn't make it back to the hive.

For the beekeeper, the first sign of a missing queen is little or no brood, no eggs, no larvae and a reduced population. Scattered drone brood where you would expect to see normal, compact brood also indicates a failed queen. When any of these occurs, the hive is in danger of collapse and urgent action must be taken.

There is a possibility that a hive with no visible brood, eggs or larvae could have a just-mated or virgin queen so look closely at the brood frames: is there honey where you would expect to see brood? The bees will often store nectar there and clean it out once a queen arrives, so nectar in brood comb is a sign there's no queen. Another test is to put a frame with eggs and young larvae from a healthy hive in the hive, making sure you shake all the bees in the frame into the donor hive first. If they start making queen cells then there is no queen.

You could let the girls continue with those queen cells but your hive numbers will dwindle substantially as it takes 14–16 days to raise a queen and then, say, 10 days for her to mate and become mature, plus another 21 days before the new brood start to hatch. That's a total of almost two months.

The quickest option is to purchase a queen. She will arrive by mail with escort bees (the queen can't feed herself so the escorts are provided to feed and attend to her needs), and will be mated and ready to go. Once she arrives keep her in a cool place (not too cold), away from ants and insecticides, sprinkling some water on the screen so she and her attendants can drink if need be until you are ready to introduce her, which should be as soon as possible.

Every beekeeper has their favourite method of queen introduction and most methods work 90 per cent of the time. If the hive is already queenless, simply take the queen cage and remove the plug or tape covering the candy plug in the cage entrance. Position the cage with the screen down on a slight angle and the candy plug up; the idea is that if the candy melts it will fall through the screen and if an escort dies it will fall away from the candy plug and not clog the exit. Push the cage into the top of some brood comb sandwiched between two frames, then close the hive up and leave it for seven days. The bees will eat out the candy plug and get used to the queen's scent at the same time; by the time she is released they will be old friends. Never release the queen directly into the hive or she will most likely be killed even if the hive is queenless.

If the hive has a queen that is laying too many drones or is ailing, your first job is to find her, which can be difficult for the novice beekeeper so seek the assistance of a more experienced hand. Once she is found you should kill her. As unpleasant as it is, this can be done by squashing her; some beekeepers believe this allows the hive to be aware that she is dead. Another way to kill the queen is to freeze her.

135

Once the queen is dead, you should wait 24 hours before introducing the new queen; this gives the bees time to realise they are queenless but not so much time that they start to raise their own queens. Leave the hive alone for seven days so the queen has time to be accepted. It's worth checking for queen cells—see page 85, Swarming—after that seven-day period and destroying them just in case: you don't want your new queen flying away in a swarm.

137

The Ullman family

Kate, Brendon, Indi, Jarrah and Pepper live on an organic farm in Daylesford, Victoria, with 1000 fruit trees in need of pollinating. They used to rent beehives from a local beekeeper but a few years ago they decided to buy their own hives and are now a family of passionate beekeepers. Each member of the family has their own bee suit and they all get involved with the whole process. They love their bees.

11

BEE HEALTH
AND DISEASES

Bees are really in trouble. We have slowly spread bee diseases all around the world, the worst being varroa mite and colony collapse disorder. Varroa has been devastating for bees as the mite carries disease and weakens colonies to the point of collapse. Colony collapse disorder (CCD) is still not understood but it is suspected pesticide use plays a big part in its impact on hives. Together, varroa mite and CCD have been responsible for devastating bee losses across Europe and the United States. At the time of writing neither disease is present in Australia, but the experts say Australia will have both soon.

As a beekeeper it is your responsibility to look after your charges and help them maintain a healthy hive. In my time as a beekeeper I have been guilty of complacency and bad practice and have suffered the consequences, with disease that has thankfully been contained.

Bees maintain a scrupulously clean environment inside their hive and then along comes the beekeeper with a dirty, contaminated hive tool, or boxes and combs from other hives, spreading disease. You need to think hospital clean when handling beekeeping equipment—for your health and for the health of the bees.

Previous page: Bees cleaning the tops of the frames.

Next time somebody offers you some old bee gear, before you say yes and delight at the savings you are going to make, consider having it irradiated to eliminate any possibility of disease. And melt the combs down, don't re-use the old wax. Irradiating is performed by specialist companies that irradiate all sorts of equipment with gamma rays, which kill all living organisms and fully sterilise the equipment.

If somebody offers you a hive, do a full inspection before you take it and consider keeping it away from your other hives for a few months until you can be sure it's clean and not likely to infect your whole apiary. Let me tell you about one time when I didn't do this and the hive was positive for American foulbrood (AFB).

I was contacted by a friend of a friend who had a hive they needed to move from their backyard. I went over there on a really wet day assuming all the girls would be inside. I had no sooner walked up to the hive in the rain than out they came and stung me a few times. OK, I thought, a touchy hive. It turned out they were not just touchy, they were plain nasty and the owner had not inspected the brood because he could not get near it. So we moved the hive one evening, getting countless stings loading the thing in the car as lots of girls were outside. In the process we also managed to forget to strap the hive down and it toppled over (but thankfully didn't separate) in the back of the car.

A week later I inspected the hive but once again the bees were so nasty I couldn't get a good look at the brood. I did manage to find the queen and kill her so it could be re-queened. I also harvested the honey and put another super on, a fatal mistake because as it turned out those frames I removed were infected with AFB.

The queen didn't take so I swapped a frame of brood out from an adjacent hive, again potentially spreading the AFB to that hive. After a successful queen, raising the hive never really took off and as the numbers dropped I was able to get a good look at the brood and diagnosed AFB.

I subsequently killed the bees in the hive using petrol fumes, done by sealing the bees in at night and squirting about 250 ml (1 cup) of petrol on top of the frames. The fumes suffocate the bees . . . even though it was necessary, I found this job quite unpleasant. All the hive equipment was then irradiated after burning the frames and dead bees. Thankfully we had good hive

143

logbooks and could identify where the frames had gone to, and so were able to keep a close eye on the hives that received infected material. It's been a year and no other AFB has materialised in those hives so it looks like we dodged a bullet.

The lessons learned were: never move frames, including stickies, from one hive to another; always log everything; and be really suspicious of any foreign hives.

Bee diseases and their treatment and regulation are rapidly changing due to the attention that has been given to bee losses in recent years. While this section can be used for information I recommend talking to your local club to find out the best treatment options for disease in your area as local beekeepers or authorities will be able to supply the most up-to-date information. The internet also has a wealth of information regarding bee diseases and their treatment.

Wax moth

The greater wax moth (*Galleria mellonella*) and lesser wax moth (*Achroia grisella*) are pests that destroy stored combs and equipment. They not only eat wax but will also burrow into woodware during their larvae stage.

Almost all beginner beekeepers make the mistake of storing equipment without treatment, only to find the comb covered in the classic dry webbing of wax moth and the woodware weakened by the burrowing larvae. Also, once you get wax moth in your house they seem to be there forever, much to the annoyance of all occupants.

Avoidance and treatment

In a warm climate, winter never seems to affect the wax moth much and it's often during winter that they'll do their worst work on those supers you have stored in your garage. This is one of the reasons that I have started leaving my empty supers on the hives over winter, with a board that has access holes cut in the corners between the hive and the super. It works for me but we have very warm winters.

I have heard of beekeepers using mothballs to keep moths at bay in stored equipment. People seem to forget that mothballs are carcinogenic and have no place in the food chain. Remember, beekeeping equipment comes into direct contact with honey.

You can try freezing equipment for 24 hours and then store it in airtight containers. I have had success with this method. I then leave the containers in an area that gets plenty of light, which the moths don't like.

Equipment to be stored can be preserved with aluminium phosphide (phosphine), but it is extremely dangerous: it's toxic to humans as well as insects and other mammals and should not be used without safety equipment. Phosphine is also used to keep beekeeping equipment free from beetles.

Wax moth is part of the life cycle of a hive and will not actually destroy an active or strong hive. So don't fear them, just be aware of them.

Small hive beetle

The small hive beetle (*Aethina tumida*) is an inconvenient critter that has changed beekeeping practices in many parts of the world forever. This beetle has also recently started attacking native hives of *Tetragonula* bees.

When I started beekeeping, people told me about this beetle and how it destroyed hives that were otherwise healthy. My belief now with some experience is that the hive needs to have low numbers and be stressed before the beetle will destroy it, but when it does, boy what a mess.

I no longer fear the small hive beetle and keep strong hives that are less prone to damage, but I am always monitoring beetle numbers and will insert traps if I feel the numbers are out of control.

The beetles carry yeasts that cause the combs they burrow through to turn into a black, smelly, festering mess. The first time I saw lots of beetle larvae I thought they were bee larvae curled up in the cells there were so many of them. Once I had cleaned up the hive there was practically a saucepan-full of the larvae and lots of stinky comb.

Originally from South Africa, the small hive beetle spread to Florida in the United States in 1998, where it caused major damage to the apiary industry. It was first discovered in Australia in 2002 and is now common in the eastern states, worse in some areas than others. The climate in eastern Australia is similar to

Florida, hence the similar damage by this pest, which thrives in humid environments.

The beetles vary in size with age but are 4–7 millimetres (⅛–¼ inch) long and 2.5–3.5 millimetres (¹⁄₁₆–⅛ inch) wide and vary from dark brown to almost black, with distinctly shaped, clubbed antennae. Their larvae are creamy white and grow to about 1 centimetre (½ inch) in length and 1.5 millimetres (¹⁄₁₆ inch) wide. They are similar to wax moth larvae but are harder to squash, with six prolegs and a row of spines along their back.

Once you meet a hive beetle you will know them. Unfortunately the bees can't kill them; they can only herd them, so they need you to kill them. Squashing them is quite satisfying.

Because the adult beetles can fly a long way (up to 10 kilometres/6 miles), seeking out beehives by smell, they can easily use feral bee colonies to spread, jumping from colony to colony.

Once in the hive, the female can lay many eggs. These are laid in irregular clusters in the comb or in crevices. The egg is approximately half the size of a honey bee egg, so really, really tiny. The eggs are laid near food sources, and to give you an idea of how many eggs: in a laboratory test, 80 beetles turned into 36,000 beetles in two months.

They hatch into larvae that burrow through the wax combs eating honey and pollen, which causes the combs to ferment and creates a slimy mess.

Mature larvae leave the hive to pupate into beetles in soil near the hive. They can crawl a long way—over 100 metres (330 feet)! So having your hive on a concrete pad doesn't help, and I have beetles in my rooftop hives with no soil in sight so that's no solution either.

Avoidance and treatment

How do you control this pest? There is no permanent solution as yet but we can use the natural instinct of the beetles to hide from bees and light to keep the numbers down.

Beetles can be hard to see if the infestation is light; you will often find them hiding under the mat when you lift it. They naturally run from light so place your super on the upturned lid.

Opposite: Placing the hive in a chicken pen is a great option for managing the small hive beetle.

After a few minutes you will find them hiding in the lid when you lift the super off.

The methods of control involve many sorts of traps used at the top of the hive or cut into the bottom board. They are all based on bees chasing the beetles so they seek refuge in the traps, which kill the beetles by drowning them in vegetable oil or soapy water, or desiccating them in diatomaceous earth. There is a trap available and registered in Australia called Apithor that uses a very small amount of insecticide that has been tested to show no transference of the insecticide into the honey.

These really annoying creatures will also damage honey and sticky frames, often within a few days. To avoid that, I freeze honey and/or stickies for two days to preserve them and kill the eggs, larvae and beetles, then seal them in airtight containers.

As with wax moths, equipment for storage can be preserved with aluminium phosphide (phosphine) but it is toxic to humans as well as insects and other mammals and should not be used without safety equipment.

I just hate these beetles and I will never forget the horror of opening a weak hive to find it completely slimed. Avoid this by keeping strong colonies and if you have a weak hive, reduce the size of the hive by removing unnecessary supers and let it grow slowly, increasing the frames only as needed so the bees can keep the beetles under control.

Nosema

Nosema is the most widespread bee disease in the world. It's a spore-based fungal microorganism that survives in the gut of bees and is spread by contaminated equipment, food, water and the bees themselves.

The easiest nosema symptom to recognise is dysentery, with faeces visible in streaks up the front of the hive and inside on the combs. This is an otherwise rare occurrence as bees are very clean and will not defecate in the hive. Unfortunately, it's easy to miss these signs and other maladies are blamed for the colony's weakness.

If the colony is infected with nosema, there is a rapid drop in its population as all bee life spans are more than halved. Young bees assume the role of the older field bees, leaving few bees to maintain the brood so bee numbers drop. In addition, the hypopharyngeal glands of infected nurse bees do not fully develop, resulting in the incapacity to properly care for developing eggs because their ability to produce royal jelly is impaired.

In serious cases the colony may die.

Avoidance and treatment

Protect colonies from cold, wet winds and locate apiaries in sheltered locations with maximum exposure to sunlight during autumn, winter and early spring. Keep them elevated and dry.

Make sure colonies have young productive queens with strong populations, especially going into autumn, and that hives have good stores of pollen and honey for winter, which encourages plentiful populations of bees.

Replace old, dark combs. This can lower the number of spores present and should be part of your hive management (see Swapping out old frames on page 83 for more information).

Research has shown that placing an infected hive in a sunny location will greatly assist in reducing the level of infection in an affected hive.

149

American foulbrood

American foulbrood (AFB) is one heck of a scary disease. It's the beekeeper's equivalent of an STD and many beekeepers don't tell anybody for fear of being targeted as a bad beekeeper. When we got AFB I chose the alternative and told everyone: the only way we can stop or at least slow it down is by spreading the word.

AFB is caused by the bacterium *Paenibacillus larvae*. It is the most serious brood disease in Australia, where it is a notifiable disease, meaning you have a legal obligation to notify authorities if your hive is infected—your local bee club can tell you who to contact. All infected hives must be destroyed and the equipment burned or irradiated by the beekeeper, which causes a significant loss to the beekeeper.

Early diagnosis of AFB will help prevent it being spread to adjacent hives and reduce the number of infected colonies. It can be spread by bees drifting from one hive to another, feeding non-irradiated honey or pollen back to bees, and infected equipment being used on different hives.

Bees stealing honey from infected hives, and beekeepers' storage areas and extraction sites also spread the disease. Specific sources of infection can include another nearby apiary or feral hive, equipment moved to the apiary or a caught swarm.

Brood should be thoroughly examined for AFB at least twice a year, in spring and autumn. It is also good practice at any time of year to examine colonies that undergo a population decline, to determine if disease is the cause.

To make matters worse, the population of an infected colony may not noticeably decrease and only a few dead larvae or pupae may be present. The disease may not develop to the critical stage where it seriously weakens the colony until the following year, or it may advance rapidly and weaken or kill the colony in the first season.

If your hive doesn't seem to be building as you would expect, it's important to inspect it for AFB. Instant test kits are now available if you don't feel confident assessing the hive yourself, or ask a more experienced beekeeper to have a look.

To inspect a hive you need to enter the brood chamber and inspect each brood frame after you have shaken the bees off, leaving the comb clearly visible for examination. The scale of dried, infected larvae are really only visible by looking from the top down into the cell with light over your shoulder shining into the base of the cells.

Also look for very dark or purple–brown brood cell cappings with a greasy appearance and some perforated cells where the bees have opened the cells in an attempt to clear them of the dead larvae. If you use a matchstick and push it into a cell that contains a dead larva it will curl out in a very characteristic rope. (Be sure to dispose of this stick into the hive after this test.)

Heavily infected hives have brood with a scattered, uneven pattern due to the intermingling of healthy cells with uncapped cells, and capped cells of dead brood with punctured and sunken cappings. This 'peppered' appearance of the brood usually allows AFB to be distinguished from European foulbrood (EFB): when infected with AFB, the cappings are discoloured, while an EFB infection results in little discolouration.

Opposite (clockwise from top left): AFB scale; a frame showing nosema infection; testing for AFB; a small hive beetle.

Avoidance and treatment

If AFB is present the colony must be destroyed before other hives get infected. Your local government body will have rules about treating AFB-infected equipment and bees. Do not treat with antibiotics as this just masks the problem while allowing the infection to spread.

Because of the risk of spreading AFB it's really important to isolate one hive's equipment from another's. Do not move stickies or hive bodies between hives after harvesting honey; effective barrier control is essential in preventing the spread of this disease.

Swarms can also carry AFB. One way of reducing this risk is to only give swarms undrawn foundation to work with. This means they must consume any honey they are carrying and the spores with it rather than storing it in comb, had drawn combs been provided.

European foulbrood

European foulbrood is just as serious as AFB and is very similar in its appearance and transmission. It is caused by the bacterium *Melissococcus pluton* and is just as easily spread as AFB by moving equipment between hives. The rope test does not work with EFB and the best method of diagnosis is to submit a slide smeared with larvae to a testing laboratory to identify the disease. Check with the laboratory for specific testing requirements. EFB can be treated with antibiotics but many beekeepers prefer not to introduce such substances into their hives and instead use the following practices: replace old queens with young, strong ones; replace old brood combs as they can harbour disease; and minimise stress—make sure the hives have sufficient food, don't shift hives unnecessarily and always at night with closed entrances.

Chalkbrood

Chalkbrood is caused by the fungus *Ascosphaera apis* and it affects both sealed and unsealed brood. The first thing you will notice is little white pellets of mummified brood at the front of the hive, deposited by the house bees cleaning out the cells.

When looking at the brood combs, larvae are at first covered with a fluffy fungus, then dry out to become hard, white or grey–

black chalk-like mummies. These mummies are deposited on the bottom board to be ejected at the front of the hive. There is no mistaking the appearance and consistency of larvae affected by chalkbrood.

Spores are highly infectious and remain viable for up to 15 years. They are brought into the hive by foraging bees, which become infected by spores left at floral and water sites by other infected bees. The infection can be spread by equipment interchange and is another reason why combs and boxes should not be moved between hives.

Avoidance and treatment

Stress of any kind—high or low temperatures, poor nutrition, relocating hives, failing queens and poor beekeeping practices—can cause the dormant disease to become apparent.

As with many diseases of bees, using new, clean wax and changing out old combs regularly can greatly assist in minimising outbreaks, as well as moving a hive to a warm place with adequate sun and better ventilation.

Placing a banana peel on top of the bars of an infected hive is also meant to help. There are various reasons why this is supposed to work, from triggering hygienic behaviours to ethylene released by the banana. I've never put the banana peel method to the test but it's worth a shot.

12

HONEY RECIPES

and ideas for using beeswax

Toasted Honey Granola

Serves 8–10

---

2 tablespoons sunflower oil

125 ml (4 fl oz/½ cup) maple syrup

2 tablespoons honey

1 teaspoon natural vanilla extract

300 g (10½ oz) rolled (porridge) oats

100 g (3½ oz) whole almonds

50 g (1¾ oz) sunflower seeds

50 g (1¾ oz) pepitas (pumpkin seeds)

100 g (3½ oz) sultanas
  (golden raisins)

30 g (1 oz) puffed millet

milk, yoghurt and blueberries,
  to serve

Preheat oven to 170°C (325°F/Gas 3).

Mix the oil, maple syrup, honey and vanilla in a large bowl. Add oats and almonds and mix well until combined.

Spread evenly over 2 baking trays and bake for 25 minutes until golden and crisp.

Remove from the oven and allow to cool. Once cool, toss through seeds, sultanas and puffed millet. Serve with milk, yoghurt and blueberries.

Store in an airtight container for up to 1 month.

Note: Puffed millet is available from health food stores.

Preparation: 10 minutes

Cooking: 25 minutes

159

Previous page: Fully capped frame of honey ready for harvesting.

Poached Rhubarb, Ricotta and Honeycomb
SERVES 8

---

530 g (1 lb 3 oz/4¼ cups) rhubarb, washed and cut into 4 cm (1½ inch) lengths

350 g (12 oz) caster sugar

90 ml (3¼ fl oz) water

½ teaspoon grenadine (optional)

200 g (7 oz) hazelnuts

500 g (1 lb 2 oz) ricotta cheese

400 g (14 oz) honeycomb

Preheat oven to 180°C (350°F/Gas 4).

Place rhubarb in a single layer in a deep-sided baking dish.

Heat sugar and water in a medium saucepan over low heat until sugar dissolves. Increase heat and bring to the boil without stirring. Pour syrup over the rhubarb and add grenadine, if using. Cover the baking dish tightly with foil and bake for 20 minutes until rhubarb is tender but still holds together. Set aside to cool. Use a spatula to transfer rhubarb to a plate. Cover with plastic wrap until ready to serve. Pour off syrup into a saucepan and reduce for 3–4 minutes until thickened to desired consistency.

Spread hazelnuts on a baking tray and toast in the oven for 5–8 minutes until golden and aromatic. While still warm, tip nuts into a clean tea towel (dish towel) and rub together to remove the skins. Chop coarsely.

To serve, divide ricotta among 8 shallow serving bowls. Add a few pieces of rhubarb and drizzle with a little reduced syrup. Portion the honeycomb into 8 and place on the ricotta and rhubarb. Scatter with the hazelnuts.

Note: Buy fresh ricotta from a cheese shop or the deli counter at a supermarket.

Preparation: 15 minutes

Cooking: 40 minutes

PANCAKES WITH HONEY BUTTER

SERVES 4

---

185 g (6½ oz/1½cups) self-raising flour

2 tablespoons sugar

pinch salt

2 eggs, lightly beaten

250 ml (9 fl oz/1 cup) milk

60 g (2½ oz/¼ cup) melted butter, plus extra, for bruhing

HONEY BUTTER

175 g (6 oz/½ cup) honey

125 g (4½oz/½ cup) unsalted butter, diced

To make the honey butter, place the honey and butter in a food processor and process until smooth. Scrape into a serving bowl and refrigerate while making pancakes.

To make the pancakes, sift the flour, sugar and salt into a bowl and make a well in the centre. Combine the eggs, milk and melted butter in a bowl and pour into the well. Whisk to form a smooth batter. Cover and allow to stand for 20 minutes.

Heat a non-stick frying pan (skillet) and brush with extra melted butter. Add 3 tablespoons of the batter to the pan and swirl gently. Cook over low heat for 1 minute, or until bubbles burst on the surface.

Turn the pancake over and cook the other side. Transfer to a plate and keep warm while cooking the remaining batter.

Spread the honey butter onto warm pancakes. Yummy.

Preparation: 5 minutes plus 20 minutes resting time

Cooking: 10 minutes

CORNFLAKE CRACKLES

SERVES 12

115 g (4 oz/⅓ cup) honey

100 g (3½ oz) butter, diced

180 g (6 cups) cornflakes

12 paper cupcake liners

Preheat oven to 170°C (325°F/Gas 3). Place 12 cupcake liners in a 12-hole (60 ml/⅓ cup capacity) muffin tin.

Mix honey and butter in a large saucepan over high heat until butter starts to froth. Add cornflakes, mix well and remove from the heat. Spoon into cupcake liners.

Bake for 10 minutes, then cool. Refrigerate until cold before serving.

Preparation: 10 minutes

Cooking: 15 minutes

HONEY ANZAC BISCUITS

SERVES 20

150 g (5½ oz/1 cup) plain flour

100 g (3½ oz/½ cup) very lightly packed brown sugar

90 g (3¼ oz/1 cup) desiccated (shredded) coconut

95 g (3¼ oz/1 cup) rolled (porridge) oats

125 g (4½ oz/½ cup) butter

115 g (4 oz/⅓ cup) honey

1 teaspoon bicarbonate of soda (baking soda)

Preheat oven to 160°C (315°F/Gas 2–3). Line 2 baking trays with baking paper.

Sift flour into a large bowl. Add brown sugar, coconut and oats and combine.

In a small saucepan heat butter and honey over medium heat, stirring until the butter melts. Remove from the heat.

Mix bicarbonate of soda with 1 tablespoon of boiling water, add to honey–butter mixture and stir in. Pour into dry ingredients and combine. Roll level tablespoons of the mixture into balls and place 5 cm (2 inches) apart on prepared baking trays. Flatten each ball with damp fingers to 1 cm (½ inch) high.

Bake for 12–13 minutes until deep golden brown, swapping the trays halfway through the cooking time to cook evenly. Leave to cool on the trays for 5 minutes before transferring to wire racks. Once cool, store in airtight containers.

Preparation: 25 minutes

Cooking: 20 minutes

French honey loaf
serves 12

---

225 g (8 oz) honey

225 g (8 oz) caster sugar

1 teaspoon bicarbonate of soda (baking soda)

250 ml (1 cup) boiling water

450 g (1 lb/3 cups) plain flour

2 teaspoons ground ginger

pinch salt

butter, to serve

Preheat oven to 120°C (235°F/Gas ½). Grease and line a 10 × 21 cm (4 × 8¼ inch) loaf (bar) tin with baking paper. Set your oven shelf to the lower third of the oven so the top of the loaf won't overbrown.

Place honey and sugar in the bowl of an electric mixer with a paddle attachment. Mix bicarbonate of soda and boiling water in a jug and pour in. Mix on low speed until the honey dissolves. Sift flour, ground ginger and salt into a bowl and add to mixer. Mix on low speed until flour is incorporated, then mix on medium speed for 1 minute until it becomes a smooth batter.

Bake for 2½ hours until the centre springs back when gently pressed.

To serve, slice thinly and spread with butter.

The loaf keeps for 5 days in an airtight container. If hot or humid, store in the refrigerator.

Preparation: 10 minutes

Cooking: 2½ hours

Bees knees cocktail

Serves 1

Honey syrup
60 g (2¼ oz) honey
60 ml (2 fl oz/¼ cup) water

60 ml (2 fl oz/¼ cup) gin or
    spirit of your choice
30 ml (1 fl oz) honey syrup
1 tablespoon lemon juice
1 tablespoon orange juice
ice, to serve

For the honey syrup, place honey and water in a small saucepan over low heat. Stir until combined. Allow to cool.

Mix gin, honey syrup, lemon and orange juice in a cocktail shaker with ice, then pour into a cocktail glass and enjoy.

Preparation: 5 minutes, plus cooling

Cooking: 5 minutes

169

Honey lemon iced tea

MAKES 1.8 litres (60 fl oz)

1.5 litres (52 fl oz/6 cups) boiling water
black tea bags
zest and juice of 2 lemons

235 g (8½ oz/⅔ cup) honey
ice, to serve

Use the boiling water to make black tea to your preferred strength. Stir in lemon zest and honey and let cool to room temperature.

Add the lemon juice, strain into a serving jug filled with ice and enjoy.

Preparation: 5 minutes

CAPPINGS AND CHEESE
SERVES 8–10

---

1 sheet honey cappings

1 baguette, thinly sliced

butter, for spreading

strong cheddar cheese,
   blue cheese or camembert,
   to serve

grapes, to serve

When extracting honey, keep a sheet of the cappings intact.

Toast thin slices of the baguette, butter each slice lightly then place a slice of cheddar, blue cheese or camembert on top.

Break off a piece of cappings and place it honey-side up on top of the cheese. Serve as a starter along with various cheeses and grapes.

Preparation: 15 minutes

Honey Mustard

MAKES ABOUT 60 ml (2 fl oz/¼ cup)

---

1 tablespoon honey

1 tablespoon hot mustard powder

2 tablespoons cider vinegar

Combine hot mustard powder and honey, then add small amounts of cider vinegar until you get the consistency of pouring cream.

This mustard is amazing with ham off the bone or on ham sandwiches, or really any other way you would normally use mustard.

Honey Salad Dressing

MAKES 170 ml (5½ fl oz/⅔ cup)

---

2 teaspoons Dijon mustard

1 tablespoon honey

80 ml (2½ fl oz/⅓ cup) Australian extra virgin olive oil

1 tablespoon sesame oil

2 tablespoons white wine vinegar

Place mustard, honey and a pinch of salt and pepper in a 200 ml (7 fl oz) jar. Mix into a paste.

Add oils and vinegar, close the lid tightly and shake. Taste and add more vinegar, mustard or oil to suit the salad.

Pour dressing over the salad just before serving.

Preparation: 5 minutes

Beetroot Carpaccio and Goat's Cheese Croquette

Serves 4

---

1 large beetroot

200 g (7 oz) goat's cheese log

1 tablespoon honey,
    plus extra to drizzle

1 teaspoon garlic oil

80 g (2¾ oz) kataifi pastry

oil spray

10 g (¼ oz) pine nuts, to serve

1 teaspoon finely snipped chives,
    to serve

2 teaspoons balsamic glaze, to serve

Preheat oven to 180°C (350°F/Gas 4).

Wrap beetroot in foil, place on a baking tray and bake for 40 minutes until tender when pierced with a sharp knife.

Cool the beetroot, peel and slice thinly using a meat slicer or mandolin to 2 mm (¹⁄₁₆ inch) thick.

To make the croquettes, mix the goat's cheese with honey and garlic oil, then roll into 50 g (1¾ oz) balls.

Loosen kataifi pastry and break off four 20 g (¾ oz) pieces. Spread each piece out into a long strip and wrap goat's cheese with kataifi pastry. Place on a baking tray and spray liberally with oil spray. Bake for 5–10 minutes until crisp and golden.

To serve, place 5 beetroot slices on each plate, overlapping each other in a circle. Top with a croquette and place pine nuts and chives on the plate. Drizzle with balsamic glaze.

Note: You can make your own garlic oil by heating 2 whole peeled garlic cloves in 125 ml (4 fl oz/½ cup) vegetable or olive oil until warm in a small saucepan. Leave to infuse for 20 minutes and strain. Keep refrigerated and discard after 1 week.

Preparation: 15 minutes

Cooking: 40 minute

Honey-Glazed Chicken Wings
Serves 4

1.5 kg (3 lb 5 oz) chicken wings

260 g (9¼ oz/¾ cup) honey

2 tablespoons soy sauce

60 ml (2 fl oz/¼ cup) tomato sauce

1 tablespoon sweet chilli sauce

2 garlic cloves, chopped very fine or minced

1 knob ginger, finely grated

1 tablespoon vegetable oil

Preheat oven to 220°C (400°F/Gas 6).

Separate chicken wings into drumettes and wings. Freeze the wing tips to make stock at a later date or discard.

To make the glaze, place honey, soy sauce, tomato sauce, sweet chilli sauce, garlic and ginger in a large jug and stir to combine.

Place chicken in a large deep-sided baking dish, add oil and toss to coat. Bake for 30–40 minutes until the skin is golden and starting to crisp. Remove from the oven and pour over the glaze, stirring to coat evenly. Return to the oven for another 10–15 minutes until dark and sticky.

Remove from the oven and let cool until the chicken can be handled. Drizzle over a bit of extra honey, season with salt to taste and serve, either warm or cold.

Preparation: 15 minutes

Cooking: 55 minutes

HONEY-GLAZED CARROTS

SERVES 6

---

3 bunches white, purple and orange
   Dutch carrots, trimmed to 3 cm
   (1¼ inch) pieces

1 tablespoon extra virgin olive oil

4 thyme sprigs

115 g (4 oz/⅓ cup) honey

Preheat oven to 220°C (400°F/Gas 6).

Place carrots in a large bowl and toss
with oil, thyme, salt and pepper. Place in a
roasting tray and bake for 30 minutes until
golden and starting to crisp.

Drizzle with honey and bake for a further
5–10 minutes until sticky and golden.

Preparation: 10 minutes

Cooking: 40 minutes

179

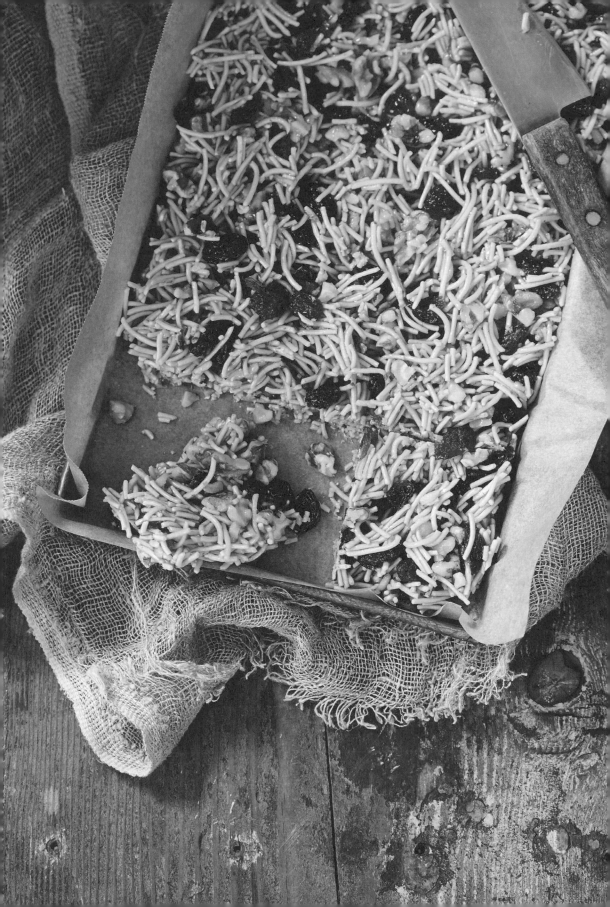

Fried honey noodles

SERVES 12

250 g (9 oz) honey

2 packets fried noodles

125 g (4½ oz/1 cup) crushed walnuts

170 g (1 cup) raisins

Heat honey in a small saucepan over medium heat until hot but not boiling.

Mix noodles, walnuts and raisins in a large mixing bowl. Pour over hot honey and mix with a spatula or metal spoon until coated.

Press mixture into a shallow baking tray lined with baking paper and refrigerate until cold. Slice into squares and serve.

Preparation: 5 minutes

Cooking: 5 minutes

CHAMOMILE HONEY ICE CREAM
MAKES 1 litre (35 fl oz/4 cups)

560 ml (19¼ fl oz/2¼ cups) milk

400 ml (14 fl oz) thin (pouring) cream (35% milk fat)

1 tablespoon chamomile tea leaves

200 g (7 oz) honey

8 egg yolks

2 cups ice, for cooling custard

waffle cones, to serve

Place milk, cream and chamomile tea leaves in a small saucepan. Bring to the boil and remove from the heat. Cover and leave to infuse for 30 minutes. Pour through a fine sieve into a large jug and discard chamomile leaves.

Heat honey in a medium saucepan over medium heat and allow to boil for 5 minutes until it begins to caramelise—do not allow it to burn. Slowly pour in milk mixture, whisking constantly until honey is dissolved.

Whisk egg yolks in a large bowl and slowly pour in hot milk–honey mixture while whisking.

Prepare a large metal bowl set over a bowl of ice—you will use this to cool the milk and yolk mixture.

Pour the milk and yolk mixture into a clean saucepan and stir over low heat until it reaches 82°C (180°F) or begins to thicken—do not allow it to boil. Pour through a fine sieve into the large metal bowl set over ice and allow to cool. Cover with plastic wrap and refrigerate until cold.

Churn using an ice cream machine according to manufacturer's instructions.

Note: Churning the mixture before it's cold may mean your ice cream doesn't set, so don't rush it. You can place the custard in the freezer for 30 minutes or so to speed up the cooling process.

Preparation: 15 minutes, plus 30 minutes infusing, 2½ hours cooling, 40 minutes churning

Cooking: 20 minutes

Bee Sting Cake

SERVES 12

Filling

500 g (1 lb 2 oz) cream cheese, softened

250 g (9 oz) ricotta cheese

zest of 2 lemons

2 tablespoons lemon juice

125 g (4½ oz/1 cup) icing sugar, sifted

Cake

300 g (10½ oz/2 cups) plain flour

55 g (2 oz/¼ cup) caster sugar

2 teaspoons dried yeast

185 ml (6 fl oz/¾ cup) milk, at room temperature

2 eggs, at room temperature

60 g (2¼ oz/¼ cup) softened butter, diced

½ teaspoon salt

Honey almond topping

90 g (3¼ oz/⅓ cup) butter, diced

50 g (1¾ oz) honey

75 g (⅓ cup) caster sugar

40 ml (1¼ fl oz) thick (double) cream

150 g (5½ oz) flaked almonds

Preheat oven to 170°C (325°F/Gas 3). Grease and line the base and side of a 23 cm (9 inch) spring-form cake tin.

For the filling, place ingredients in an electric mixer fitted with a flat paddle and mix until just combined. Refrigerate until needed.

For the cake, place all ingredients in the bowl of an electric mixer and mix with a flat paddle attachment on medium speed for 2 minutes. Scrape the batter into the spring-form cake tin and spread out evenly with a spoon or offset spatula. Place the tin in a clean plastic bag and tie to enclose. Leave in a warm, draught-free place and allow to rise for 1–1½ hours.

Meanwhile, make the topping. Place butter, honey, sugar and cream in a medium saucepan over medium heat and bring to the boil. Simmer for 5–6 minutes until the mixture turns a slightly darker shade of yellow. Add almonds and stir through.

>>>>→

Remove from the heat and allow to cool until the cake dough has finished rising.

Starting at the edges, spoon the topping evenly over the cake dough and bake for 30 minutes until golden brown. Cool in the tin for 10 minutes on a wire rack set over a baking tray to catch any caramel drips. Release the side of the cake tin and remove the baking paper from the side. Leave until completely cool before slicing.

Use a long serrated knife to halve the cake horizontally. Place the top half on a flat plate.

Spread the filling evenly over bottom half of the cake using a palette knife or spatula and cover with the top layer of the cake.

Bee sting cakes are best eaten on the day they are baked, though they can be partly prepared the day before: keep the risen dough in an airtight container at room temperature, then bake and fill the following day.

Note: You can also use a non-spring-form tin; just let the cake cool for 20 minutes before turning it out. Be very careful as the caramel will still be warm.

Preparation: 30 minutes, plus 1–1½ hours rising time, plus 2 hours cooling

Cooking: 40 minutes

Honey milk mousse

Serves 6

---

250 ml (9 fl oz/1 cup) milk

100 g (3½ oz) honey

½ vanilla bean, seeds scraped

2½ leaves (9 g) titanium-strength gelatine, soaked in cold water for 5 minutes

250 ml (9 fl oz/1 cup) thin (pouring) cream (35% milk fat)

70 g (2½ oz) egg whites

100 g (3½ oz) caster sugar

6 fresh figs, to serve

honey, to drizzle

Place milk in a small saucepan with honey and vanilla bean and cook over medium heat until hot but not boiling. Squeeze water from soaked gelatine, add to honey–milk mixture and stir until dissolved. Strain mixture through a sieve into a large metal bowl and allow to cool. Cover the bowl with plastic wrap and refrigerate for 45 minutes. Stir and refrigerate for another 15–30 minutes until cold and starting to set around the edges.

Whip the cream until soft peaks form using hand beaters or an electric mixer with the whisk attachment. Refrigerate.

Once the honey–milk is almost set, place egg whites and sugar in the bowl of an electric mixer and whisk on high speed for 8 minutes until it forms a meringue and the sugar has dissolved. Fold the cream gently through the meringue.

Fold the almost-set honey–milk through the meringue cream and divide among 6 serving glasses or bowls. Cover and refrigerate until ready to serve.

Serve with quartered figs and a drizzle of honey.

Preparation: 30 minutes

Chilling: 1½ hours

Cooking: 10 minutes

187

Honeycomb using honey

Serves 10–12

115 g (4 oz/⅓ cup) honey

330 g (1½ cups) caster sugar

3 tablespoons water

3 teaspoons bicarbonate of soda (baking soda), sifted

Line a 3 × 22 × 33 cm (1¼ × 8½ × 13 inch) baking tray with a large sheet of baking paper that extends up the sides.

Place honey, sugar and water in a 3 litre (100 fl oz) saucepan and cook over medium heat, swirling the pan occasionally, until sugar dissolves. Increase heat to high and boil for 5–6 minutes until it starts to caramelise and becomes a few shades darker—do not allow to burn. Add bicarbonate of soda and stir briefly, then scrape and pour into the prepared tray and let cool.

To serve, break into pieces and enjoy. Store in an airtight container.

Note: If you only have a shallow baking tray handy, make half the quantity, using 2 tablespoons each of honey and water, 165 g (5¾ oz/¾ cup) caster sugar and 1½ teaspoons bicarbonate of soda (baking soda).

Preparation: 10 minutes, plus cooling

Cooking: 10 minutes

Honey-roasted macadamias

Serves 10

500 g (1 lb 2 oz) macadamia nuts

100 g (3½ oz) honey

50 g (1¾ oz) butter, diced

2 teaspoons sea salt flakes

Preheat oven to 180°C (350°F/Gas 4).

Place macadamias on a baking tray lined with baking paper and roast for 8–10 minutes until golden.

Place honey in a medium saucepan and bring to the boil. Simmer until caramelised, then add nuts and quickly stir through. Add the butter and stir over low heat until combined. Season with the

sea salt. Pour onto the lined baking tray, scraping out the saucepan with a spatula. Smooth into a flat layer and leave to cool.

Preparation: 10 minutes

Cooking: 25 minutes

Pear and honey curd

MAKES 600 ml (21 fl oz)

---

pears, enough to make 200 ml
(7 fl oz) fresh pear juice

8 egg yolks

100 g (3½ oz) honey

1 leaf titanium-strength gelatine, soaked
in cold water for 5 minutes

160 g (⅔ cup) butter, cut into cubes

Juice pears and strain through a fine sieve so you retain only the clear liquid.

Mix yolks, honey and pear juice in a wide metal bowl. Place the bowl over a large saucepan filled with 5 cm (2 inches) of gently simmering water and whisk by hand for 5–6 minutes until the mixture forms thick ribbons on itself (known as the sabayon stage).

Squeeze water from soaked gelatine and add to hot mixture, whisking until dissolved. Whisk in butter cubes a few at a time, waiting until they have been incorporated before adding the next few. Strain through a fine sieve into a container with a lid and refrigerate. If the mixture sets a little firm, whisk briefly to loosen.

Note: Pear juice can be replaced with any fruit juice or liquid that goes with honey, such as apple juice.

Preparation: 20 minutes, plus 3 hours setting time

Cooking: 25 minutes

190

Bees produce a number of substances that are collected and sold: everything from their venom, royal jelly, wax, propolis and honey. Some beekeepers also scrape the pollen from bees' legs using traps and then sell it. I do not collect pollen, royal jelly or venom as I do not agree with those practices but I do use and sell honey and wax.

Beeswax

You can tell a beekeeper's house because you will find little pellets of wax around the place. Bees produce wax in little flakes under their abdomen and shape these flakes with their mouths into the honeycomb shapes we recognise. The wax is costly for them to produce from an energy perspective. Unlike paraffin wax or soy wax, beeswax is natural, unprocessed and non-toxic. It's 100 per cent hydrocarbon and combusts with no smoke or other particulates, so it's fantastic for candle-making.

Processing comb and capping

Turning a pile of old bee combs and cappings into pure yellow beeswax takes a bit of work; the old combs and cappings contain cocoons and other non-wax components that need filtering out before you end up with the good stuff. When I had just a few hives and produced only a small amount of wax I extracted it on the stove. I wrapped the wax in an old t-shirt, put a brick on it and placed it in a big old stainless steel tank full of water with a gas ring under it. When the water was close to boiling I turned the gas down; the wax would melt out through the fabric and be filtered to float on the surface of the water. Once the water cooled, there were clean lumps of floating beeswax that could be combined, melted and turned into candles.

We now have so much wax I use a couple of processes. To melt old comb I have an old stainless steel extractor body with a filter grate in the bottom and a lid. I have a hole drilled into the side through which I pipe steam from a wallpaper stripper. I fill the unit with frames on their sides, put the lid on and when the steam generator is turned on, the steam melts the wax away from the comb and pure wax dribbles out the side. Once all the wax has melted, all that's left in the combs is cocoons that can be discarded (makes great fertiliser for the garden).

Opposite: Common equipment used by beekeepers, including protective clothing, the smoker and hive tool.

To process cappings I use a machine our bee club purchased and lends to its members. It consists of a 100-litre (210-pint) stainless steel insulated vessel with a heating element at the bottom, along with a drain valve. Just above the element position is a small water inlet with a valve and at the top a wax outlet with a simple screen.

To use the unit you fill it up with water until the element is covered. Add wax cappings until the water level is right up to the top, then put the lid on. Connect the power and turn the thermostat up to boil. Once the water has boiled and all the wax has melted, turn the thermostat down so that the water remains hot but not boiling, then leave it alone for four hours. After four hours the sediment falls out of suspension and you'll find wax floating on top. Water is very slowly introduced via the lower inlet and as the water level rises, clean wax flows out of the outlet.

It's the simplest machine but works miracles, with bright, clean, clear wax being produced—so much better than anything I have tried.

For the small-scale beekeeper you could try making a solar wax melter from a sheet of glass or an old window placed over a polystyrene box. The whole contraption is angled towards the sun and the wax can be suspended over a filter cloth that is stretched over a container or bucket. The heat will melt the wax through the cloth into the container, filtering as it goes. There are quite a few designs available online, have a look at YouTube or www.beesource.com.

Candles, balms and polishes

Beeswax has loads of different uses. You could use your wax to try making candles, lip balms and soap, or a furniture polish.

Candles

How do you turn wax into candles? That's the easy part. First, your local candle-making or beekeeping store can sell you wick; make sure you buy wick specially made for beeswax as other types of wicks will not burn correctly.

Select a mould or container for your candle (an old jar works really well), and suspend the wick in the jar with a paper clip and a wooden stirrer placed across the top of the container. Melt your wax in some old saucepans that you never want to cook in again using a double boiler process: a bottom pan full of water and sitting in it a smaller pan containing wax. Once your wax has melted, pour it into your candle mould or jar. To prevent too much shrinkage of the cooling wax I stand my jars in a pan filled with boiling water, so it cools slowly. Once it's all cool, trim the wick. And there you have it: a candle made from your own beeswax!

Lip balm
This is ridiculously easy. All you need is:
> 3 teaspoons of your own beeswax
> 5 teaspoons of castor oil (as a carrier)
> ½ teaspoon vitamin E oil
> 1 teaspoon honey
> a little jar to put it in

First, melt the beeswax and carrier oil together in a double boiler or in a heatproof bowl in the microwave. Once it's runny and combined, mix in the vitamin E and honey and pour into your jar . . . that's it.

195

Furniture polish
A lovely furniture polish can be made by combining equal parts:
> beeswax
> turpentine
> linseed oil

Heat the beeswax in a double boiler until molten, then add warm turpentine and linseed oil. You can adjust the quantity of beeswax to increase or decrease the hardness of the wax.

Once combined and cool, pour into a suitable glass jar—the oils may degrade plastic containers.

Honey as an antibiotic
The ancient Egyptians were way ahead of us: instead of putting honey in their tea or spreading it on crumpets they used it as medicine. I have seen an Egyptian hieroglyph depicting a medical procedure with a beehive nearby: honey was used for disinfecting the wound.

Modern medicine is now finally waking up and discovering how honey can be used as an antibiotic that actually promotes

healing and works where some more chemical treatments have failed. The process by which honey does this is not fully understood; in tests that compare the antibiotic action of raw honey with phenol, scientists have removed the compounds from the honey that they know are antibiotic and yet it still works.

For years humans have been using honey on cuts and scrapes. I would be cautious using honey on open wounds if it hasn't had the unwanted bugs killed by irradiation, but I have tried my honey and it does seem to work.

Propolis is also used in medicine and has been tested against cancers, though the results are yet to be fully analysed.

Making money from the honey

Some people approach beekeeping thinking it will be an easy money-maker, but nothing could be further from the truth. Sure, you can sell your honey but it's not a get-rich-quick scheme, more a get-worn-out-and-tired-quick scheme!

Beekeeping is very manual work; unless you have a large number of hives the only automatic equipment you will have will be an electric extractor. I often quip that I wish bees made marshmallows because honey is very heavy; at 1.4 times the weight of water it is very dense, and you are doing lots of heavy lifting in the summer heat.

Having said that, you can make some holiday money by selling honey at your local farmer's market; consumers are looking for local produce and honey is often the most local product at these markets.

You also need to look at the food health regulations in your area, and possibly labelling and licensing requirements, as there are usually special requirements for selling food that also apply to honey.

Beekeeping is exceptionally rewarding, and by becoming a beekeeper you are helping to support a vital insect that is being driven to extinction by our own stupidity. Once you have your first hive you will (if you are like me) spend a fair bit of time just watching the girls come and go and wondering where they have been. You could end up devoting so much time to them that your partner may start muttering about bee fever.

Opposite: A bee stealing honey from cappings.

By maintaining a hive, a beekeeper is benefitting their whole neighbourhood by providing pollination services to local flora in exchange for nectar, a symbiotic relationship that has existed since before the dinosaurs (and humankind) trod the earth.

Tell everybody about bees and what they do for us and why we need them. Together we can help save our bee populations. So get a hive and enjoy the delights of beekeeping.

GLOSSARY

bee space
Approximately 6–8 millimetres (¼–⅜ inch) of space that bees generally will not fill with brace comb or propolis. Allows the standard hive designs to work and combs to be accessed without destroying the hive.

brace comb
Comb built by bees in odd places. Brace comb is not usually used to store honey and is most often seen between the top bars of frames.

brood
Baby bees in various stages of development, from eggs to grubs to capped brood to hatching bees.

burr comb
Often seen in the lid, burr comb usually contains honey and is built where there is enough space and the bees want to use it. It can be reduced by using a hive mat.

capped brood
The stage where a tan-coloured cap is placed over a developing bee cell, after which the bee spins a cocoon and metamorphoses from a grub into a bee.

capped cells
Honeycomb cells full of honey that have been closed up with wax by the bees.

cappings
The wax left over from uncapping honey. Capping should be strained to release the honey before rendering it.

cut comb
Squares of honeycomb cut from full frames using a sharp knife or cookie-cutter-like gadget.

drawn frames
Frames with wax honeycomb structure created by the bees.

foundation
Thin embossed wax sheet used with frames to strengthen the honeycomb.

frames
A four-sided timber structure machined to exact tolerance so that a group will fit into a super and conform to bee space. Once drawn, frames allow easy honey harvesting or inspection of brood.

hive body
A timber box with sides and no lid or bottom used to hold frames and form a beehive. Can be in a number of sizes.

irradiation
Irradiating is performed by specialist companies that irradiate all sorts of equipment with gamma rays, which kill all living organisms and fully sterilises the equipment.

laid out
A brood frame that has been laid with eggs and may contain developing bees in various stages of development.

nectar flow
When flowering plants are producing nectar that is being stored in the hive to make honey. Can be assessed by shaking a honey frame on its side over the hive—if there is nectar flow, fresh nectar will fall out.

nucleus
A small beehive with four or five frames. It has many uses including for mating and as a hospital; the smaller size makes it easier for bees to manage the temperature.

propolis
Bee glue made from tree sap. A very sticky substance found gluing things together in a beehive. Propolis has many medical uses.

201

re-queening

To replace the queen in a beehive with a new one, usually done with a special cage that allows the hive to get used to the new bee scent before the replacement queen is released.

stickies

Honey frames from which the honey has been harvested. The frames remain sticky with honey and are usually placed back in the hive so the bees can clean up the honey.

supers

A timber box with sides and no lid or bottom used to hold frames and form a beehive. Can be in a number of sizes. Honey is stored in the supers.

undersuper

To move a honey super up and place an empty super under it, so that the bees continue to fill the upper super before starting on the lower empty one.

ACKNOWLEDGEMENTS

Richard Braithwaite for giving me my first hive and helping start my journey, George Schwartz for his philosophy of beekeeping, swarms and glasses of mead and Susan Burchill for her encouragement and punctuation. Vicky Brown without whom The Urban Beehive couldn't exist, Ian Phillips for his bee wrangling of my first grumpy hive and proofreading, my fellow beekeepers George and Charis, Ian, Elke, Marion and Brian, Matt and Vanessa, Tessa, Bruce and Rahni, Vicky, Katrina and Jonathan and the Ulman family for allowing us into their homes and hives. Thanks to the recipe contributors Brett Armstrong for his Toasted honey granola and Poached Rhubarb, Ricotta and Honeycomb; Elke Haege for the Honey salad dressing; Stephan Tseng for the Beetroot carpaccio and goat's cheese croquette; Adam Hall for the Chamomile honey ice cream, Honey milk mousse and Honey-roasted macadamias. Thanks to all the beekeepers who have patiently answered so many questions from me along the way. The shoot team: Cath Muscat, for her incredible photography skills getting the best out of people and bees; Michelle Noriento, for her food styling and ironing skills; and Tina Asher, for her cooking. Also designer Astred Hicks, editor Sophia Oravecz, the team at Murdoch books including Jane Morrow, Claire Grady, Megan Pigott and Hugh Ford who patiently guided me through this project. And finally, to the bees who work tirelessly pollinating our food and making honey—without them we'd be in a pickle.

202

Doug Purdie, along with his partner at The Urban Beehive, maintains more than 70 beehives on city rooftops, balconies, backyards and in community gardens around Sydney. He runs beginner beekeeping courses and is president of the Amateur Beekeepers' Association of NSW.

INDEX

Page numbers in *italics* refer to photographs.

203

204

205

Recipe Index

Page numbers in *italics* refer to photographs.

206

Published in 2014 by Murdoch Books, an imprint of Allen & Unwin.

Murdoch Books Australia
83 Alexander Street
Crows Nest NSW 2065
Phone: +61 (0) 2 8425 0100
Fax: +61 (0) 2 9906 2218
www.murdochbooks.com.au
info@murdochbooks.com.au

Murdoch Books UK
Erico House, 6th Floor
93-99 Upper Richmond Road
Putney, London SW15 2TG
Phone: +44 (0) 20 8785 5995
www.murdochbooks.co.uk
info@murdochbooks.co.uk

For Corporate Orders & Custom Publishing contact Noel Hammond,
National Business Development Manager, Murdoch Books Australia

Publisher: Jane Morrow
Photographer: Cath Muscat
Styling: Michelle Noerianto
Food editor: Tina Asher
Designer: Astred Hicks, Design Cherry
Design Managers: Hugh Ford and Megan Pigott
Editor: Sophia Oravecz
Editorial Manager: Claire Grady
Production Manager: Mary Bjelobrk

A cataloguing-in-publication entry is available from the catalogue of the
National Library of Australia at www.nla.gov.au.

A catalogue record for this book is available from the British Library.

Colour reproduction by Splitting Image, Clayton, Victoria.

Printed by Hang Tai Printing Company Limited, China.